Hypersonic Vehicles - Past, Present and Future Developments

Edited by Giuseppe Pezzella and Antonio Viviani

Published in London, United Kingdom

IntechOpen

Supporting open minds since 2005

Hypersonic Vehicles - Past, Present and Future Developments
http://dx.doi.org/10.5772/intechopen.79515
Edited by Giuseppe Pezzella and Antonio Viviani

Contributors
Sannu Mölder, R. C. Mehta, Mookesh Dhanasar, Frederick Ferguson, Julio Cesar Mendez, Yao Zheng, Lei Fu, Meijun Zhu, Shuai Zhang, Shuai Zhou, Minghui Chen, Tianlai Gu, Antonio Viviani, Andrea Aprovitola, Luigi Iuspa, Giuseppe Pezzella

First published in London, United Kingdom, 2019 by IntechOpen
IntechOpen is the global imprint of INTECHOPEN LIMITED, registered in England and Wales, registration number: 11086078, The Shard, 25th floor, 32 London Bridge Street
London, SE19SG - United Kingdom
Printed in Croatia

British Library Cataloguing-in-Publication Data
A catalogue record for this book is available from the British Library

Additional hard and PDF copies can be obtained from orders@intechopen.com

Hypersonic Vehicles - Past, Present and Future Developments
Edited by Giuseppe Pezzella and Antonio Viviani
p. cm.
Print ISBN 978-1-83962-269-4
Online ISBN 978-1-83962-270-0
eBook (PDF) ISBN 978-1-83962-271-7

For EU product safety concerns:
IN TECH d.o.o., Prolaz Marije Krucifikse Kozulić 3, 51000 Rijeka, Croatia, info@intechopen.com or visit our website at intechopen.com.

Meet the editors

Giuseppe Pezzella obtained an MSc in Aeronautical Engineering in 1999 from the University of Naples "Federico II." He obtained a PhD in Aerospace Engineering from the same university in 2003. Dr. Pezzella is a lecturer in Aircraft Aerodynamics and Aerothermodynamics at the Department of Engineering of the University of Campania "L. Vanvitelli," Italy. From 2005 to 2018 he worked as a research engineer in the Physics of Fluids Department at Centro Italiano Ricerche Aerospaziali (CIRA).
He is a specialized analyst in the fields of vehicles' aerodynamics, aerothermodynamics, and aeroshape design. He is involved in national (e.g., PRORA) and international projects (e.g., EXPERT, HIGH-LIFT, FLPP-IXV, RASTAS-SPEAR, HEXAFLY, LAPCAT-II, SCRAMSPACE, HEXAFLY-INT). He is also currently involved in NATO STO research activities. He is author or co-author for about 100 publications among archive journals, conference papers, and edited books. He is also a reviewer for several international journals

Antonio Viviani has advanced degrees in Aeronautical Engineering and Aerospace Engineering from the University of Naples. He is full Professor in the Engineering Department at the University of Campania where he is also chairman of the teaching staff in Aerospace Engineering.
Antonio is a member of AIAA, AAAS-American Association for Advancement of Sciences, ELGRA-European Low Gravity Research Association, NYAS-New York Academy of Sciences; member of the Physical Sciences Working Group of ESA-European Space Agency (1993–1996; 1996–1999); member of the Microgravity Sciences and Processes Committee of the IAF-International Astronautical Federation (1994–); member (elected) of the Management Committee of ELGRA (1995–1997; 1997–1999); Vice-President (elected) of ELGRA (1999–2001; 2001–2003); Vice-Chairman (elected) of the Committee for Microgravity Sciences and Processes of the IAF (2003–2006); Chairman (elected) of the Committee for Microgravity Sciences and Processes of the IAF (2006–2009; 2009–2012); member (permanent, elected in 2004) of the IAA- International Academy of Astronautics; member of the Commission 1 (Space Physical Sciences) of the IAA (2007–2009; 2009–2011; 2011–2013; 2013–2015); Regional Secretary (for Italy) of the IAA (2007–2009; 2009–2011; 2011–2013); member of the Awards and Memberships Committee of the IAA (2011–); Member of the Publications and Communications Committee of the IAA (2011-); and co-editor of *Acta Astronautica* (2011–).
Co-Investigator of the experiment "Onset of Oscillatory Marangoni Flows", Mission Spacelab D2 - Space-Shuttle Columbia, Flight STS-55 (April 1993). Principal Investigator of the experiment "Bubble Behavior under Low Gravity", Mission Spacelab IML-2 - Space-Shuttle "Columbia", Flight STS-65 (July 1994). Principal Investigator of the experiment "Non-linear Surface Tension Driven Bubble Migration", Mission Spacelab LMS – Space Shuttle "Columbia", Flight STS-78 (June 1996).

Contents

Preface

Hypersonic vehicles represent the next frontier of transportation systems. This is true both for travels to and from space and for fast passenger transportation. Indeed, after the retirement of NASA's Space Shuttle, scientific applications in Low Earth Orbit (LEO) strongly highlight the need for flexible, affordable, comfortable and safe routine access to space.

In the last decade, the attention to hypersonic travel for civilian application has dramatically increased. Many start-up and aerospace companies are focusing attention on the potential benefit of hypersonic aircrafts able to fly from point to point all over Earth in less than 2–3 hours for both passenger and goods transportation.

As a result, the aerospace field, despite past important achievements, still focuses greatly on high-speed vehicle design.

The quest to fly higher and faster, however, is still a challenging goal for the field of aerospace engineering, especially for human-rated missions. In fact, high-speed hypersonic vehicles demand several design issues to be addressed, such as aircraft aerodynamics and aerothermodynamics, aeroshape design optimization, aerodynamic heating, boundary layer transition, propulsion system integration, low-speed flying qualities, and so on.

Among others, aeroshape optimization perhaps represents the fundamental aspect in designing such vehicles. Indeed, high-speed vehicle design is an extremely challenging process involving several disciplines, including aerodynamics, aerothermodynamics, control, avionics navigation systems, propulsion, and structure. As well known, these disciplines are strongly coupled with one another and generally influence each other because they involve antagonistic objectives. Therefore, it is expected that synergistic interactions between vehicle sub-systems and functions can produce an optimized multidisciplinary vehicle design with significant performance and economic improvements.

In this framework, this book contains contributions focusing on hypervelocity aircraft design. For instance, the extreme loading environment of high-speed flows demands advanced aeroshape designs that are able to sustain large heat transfer and mechanical loading conditions while providing aerodynamic force and moments to sustain both re-entry and hypersonic flights. Therefore this book covers such topics as hypersonic aircraft aerodynamic and aerothermodynamic design, aeroshape design optimization, computational fluid dynamics, and scramjet propulsion. The book also discusses high-speed flow issues and the challenges to achieving the dream of affordable hypersonic travel. It is hoped that the information contained herein will allow for the development of safe and efficient hypersonic vehicles.

Giuseppe Pezzella and Antonio Viviani
Engineering Department,
University of Campania "L. Vanvitelli",
Aversa (CE), Italy

Section 1

Hypersonic Vehicles: Past, Present and Future

Chapter 1

Introductory Chapter: Hypersonic Vehicles - Past, Present, and Future Insights

Antonio Viviani and Giuseppe Pezzella

1. Introduction

In the aviation field, great interest is growing in high-speed vehicle design. The increase in manned and unmanned space operations in low earth orbit (LEO) demands an evolution in the vehicle for payloads transportation up to and from LEO to improve the levels of flexibility, affordability and safety of routine access-to-space missions. Today this need is utmost stringent in the light of the NASA Space Shuttle retirement.

On the other hand, in the last few years, the attention to hypersonic travels for civilian application has also increased dramatically. Many start-up industries are focusing attention on hypersonic aircrafts able to fly, e.g., from New York to Sydney in less than 2–3 hours, thus providing a lot of insights on the oncoming market of hypersonic flights.

As a result, over the decades the potential benefit of an operational hypersonic vehicle (HV) has driven continued researches in basic and applied technologies. Indeed, several high-speed aircraft concepts (i.e., lifting and winged vehicles) have been conceived or developed in the USA, Russia (former USSR), Europe, France, Germany, Italy, and Japan to simplify access to LEO and sustained high-speed flight routinely and in a safe way. Most of these projects, however, were just prototypes or developed at the conceptual design stage and linked to flight testing focused mainly on some technologies rather than assessing the effectiveness and the advantages of a cutting-edge design.

A look on the HV research programs developed so far is hereinafter described for each country.

2. Past developments

US research on HVs lies on the X-plane (XP) programs. They are a series of experimental aircrafts to test and evaluate innovative technologies and aerodynamic concepts. Most of the XP have been operated by the National Advisory Committee for Aeronautics (NACA) or, later, the National Aeronautics and Space Administration (NASA), often in conjunction with the United States Air Force (USAF). The manned Bell X-1 and the North American X-15 were the most famous [1]. The former was the first aircraft to break the sound barrier in level flight in 1947. The latter was a hypersonic rocket-powered aircraft that achieved the world record for the highest speed ever recorded by a manned vehicle at the time [1].

Later XPs which supported important research in a multitude of aerodynamic and technical fields were unmanned (some were remotely flown, some were partially or fully autonomous) such as the X-20 Dyna-Soar, X-23A, X-24, X-30, X-33, X-34, X-37, X-38, X-43, X-51, and Dream Chaser [1]. The X-20 was a USAF program to develop a spaceplane for several military missions, including satellite maintenance or sabotage and aerial reconnaissance. The vehicle configuration was innovative and more like the much later Space Shuttle. It was designed to glide to Earth like an aircraft under the pilot's control and land on runway, rather than simply falling to Earth and landing with a parachute. The program started on October 24, 1957, but was cancelled in 1963 just after spacecraft construction. The X-23A was a small lifting body tested by the USAF to study the effects of maneuvering during reentry, including cross-range maneuvers. The X-24 was developed from a joint USAF-NASA program. It was designed and built to test lifting body concepts, experimenting with the concept of unpowered reentry and landing, later used by the Space Shuttle [2]. The X-30 was an advanced technology demonstrator for the National Aero-Space Plane (NASP) program to create a single-stage-to-orbit (SSTO) spacecraft and passenger spaceliner. It started with the aim to advance and demonstrate hypersonic technologies for scramjet-powered vehicles. The X-30 was cancelled in the early 1990s before a prototype was completed, although much development work in advanced materials and aerospace design was completed [3, 4]. The X-33 was an unmanned, subscale technology demonstrator for the VentureStar orbital spaceplane, which was planned to be a next-generation reusable launch vehicle (RLV) [5, 6]. The X-33 would flight-test a range of technologies that NASA deemed critical for SSTO RLVs, such as the aerospike engine, metallic thermal protection system (TPS), and its lifting body aerodynamics. The X-34 was intended to be a low-cost testbed for demonstrating RLV key technologies [7]. It was conceived to be an autonomous pilotless craft capable of reaching Mach 8 and performing 25 test flights per year. The X-37 is a reusable HV developed as orbital test vehicle (OTV) [8, 9]. An early goal for the program was in-orbit operations for the spacecraft to rendezvous with satellites and perform repairs. The technologies demonstrated in the X-37 include an improved TPS, enhanced avionics, an autonomous guidance navigation, and control system. The X-38 was an experimental reentry vehicle designed by NASA to research a possible emergency crew return vehicle (CRV) for the International Space Station (ISS) [10].

Following NASP, in the 2000s, NASA concern was National Aerospace Initiative (NAI) where a large use of multidisciplinary design was undertaken. The NAI's mission was to ensure the USA's aerospace leadership with an integrated, capability-focused, national approach that enables high-speed/hypersonic flight; safe, responsive, affordable, reliable access to and from space; and in-space operation by developing, maturing, demonstrating, and transitioning transformational aerospace technologies. In the framework of NAI, NASA performed several in-flight validations of hypersonic technologies and evaluation of new concepts, such as X-43 and X-51 vehicles [11, 12]. The X-43 program set out to demonstrate hydrogen-fueled scramjet operations in a fully integrated aircraft system at Mach numbers of 7 and 10. The X-43 vehicle was a 4-m-long lifting body design, weighing about 1500 kg, with a fully integrated scramjet engine. The two flights of the X-43A vehicles were successful in achieving all research objectives. Comparisons to ground test in shock-heated tunnels confirm the ability of these facilities to measure engine performance consistent with flight.

The recently successful X-51 flight program went even further in technology development and mission objectives. After release from the B-52 carrier, the waverider-derived vehicle is accelerated by an ATACMS booster to Mach 4.5, whereas the scramjet engine further accelerated the vehicle up to about Mach 6. Furthermore, the engine flowpath is cooled using fuel in a cooled loop to both maintain tolerable flowpath temperatures and crack the fuel to facilitate ignition once it is injected into the combustion region of the scramjet engine. The external vehicle configuration utilizes a waverider-type aerodynamic mold line as forebody to maximize hypersonic L/D ratios. After three (partial) failures, the final flight of the X-51A Waverider test program has accomplished a breakthrough in the development of flight reaching Mach 5.1 over the Pacific Ocean on May 1. The cruiser travelled over 250 km in just over 6 minutes. It was the longest of the four X-51A test flights and the longest air-breathing hypersonic flight ever.

The Dream Chaser is a reusable lifting-body spaceplane being developed by Sierra Nevada Corporation (SNC) [13].

Russia (former USSR) involvements in developing high-speed vehicles and their related propulsion units refer to various Mach numbers. Russia was and still is very active in high-speed vehicle and propulsion design for various Mach numbers. However, limited information is available. The most important was the OK-1K1 spaceplane, referred to as Buran [14]. It was the first spaceplane to be produced as part of the Soviet/Russian RLV program and is the only Soviet RLV to be launched into space.

In Europe, in the last years, a whole range of space transportation concepts under various research and development programs have been investigated. During the 1980s and 1990s, there was significant interest in designing a RLV. National space agencies such as the Centre National d'Études Spatiales (CNES) of France and the German Aerospace Centre (DLR) worked on their own designs, the most prominent of these to emerge being the Hermes spaceplane of CNES [15]. It is very similar to the X-20 and the Space Shuttle. As intended, the Hermes was an RLV to transport both astronauts and moderate-size payloads into LEO and back again. In comparison to the Shuttle, Hermes is a substantially smaller vehicle and does not share the ogival platform of the Orbiter. But, it was designed with a highly swept delta wing with wingtips, close to the X-20. Then, Hermes was later further developed by the European Space Agency (ESA) for several years but was ultimately terminated in 1992 prior to any flights due to numerous delays for unachievable performance goals and funding issues. After the abandonment of the Hermes program, ESA, however, decided to maintain the strategic long-term objective to develop a RLV. It was started the Future European Space Transportation Investigations Programme (FESTIP). In the framework of FESTIP, the Hopper concept was envisioned by the ESA as RLV. It was one of the several concepts to function as a European RLV for the inexpensive delivery of payloads into orbit [16]. A prototype of Hopper, namely, Phoenix, was tested within the wider ASTRA program of the DLR [17, 18]. After that, ESA started the Future Launchers Preparatory Programme (FLPP) [19]. Under FLPP, Europe has undertaken detailed investigations of several partially reusable launch concepts with the aim to develop a next-generation launcher [19]. A total of four concepts were investigated, namely, the horizontal take-off (HTO) Hopper, the vertical take-off (VTO) Hopper, the reusable first stage, and the liquid fly-back booster [20]. Each of these concepts consisted of a reusable winged booster, able to carry an expendable upper stage, to deliver a payload in geostationary transfer orbit [21]. The HTO Hopper featured a relatively conventional wing-body configuration, investigated yet within the ASTRA program [17]. The VTO-Hopper was designed with a traditional slender missile-like body but with a small delta wing and a central vertical stabilizer arrangement [21]. To test and further develop the technologies and concepts produced by these studies, there was a clear need to accumulate practical flight experience. To this end, ESA has undertaken the design of the intermediate experimental vehicle (IXV) also promoted within the FLPP framework [22]. It was derived by the Pre-X concept investigated early by CNES [23]. The IXV holds the distinction of being the first ever lifting body to perform full atmospheric reentry.

German studies refer to the SHarp Edge Flight EXperiment (SHEFEX) program of DLR for the development of future reentry and hypersonic technologies [24]. The goal is to set up a flying laboratory to gain knowledge of the physics of hypersonic flow, complemented by numerical analysis and ground-based testing. SHEFEX flight experiments were an excellent laboratory to test new technological concepts and in-flight experimental sensors.

In Italy there was the unmanned space vehicle (USV) program [25]. Within the ongoing USV project, CIRA conceived a family of flying test beds (FTB's) for in-flight experiments in the fields of aerodynamics, aerothermodynamics, flight mechanics, control, and aeroelasticity. The first phase of the USV Program consisted of the design and realization of two laboratories (i.e., FTB-1). The FTB-1 concept was based on a winged slender-body vehicle able to address in-flight experiments and low atmosphere maneuvered flights at supersonic, transonic, and low subsonic Mach numbers, referred to as dropped transonic flight test (DTFT) missions. The flight test success demonstrated the ability of designing and implementing robust guidance and control laws up to low subsonic Mach numbers.

Japan contributions to unmanned RLVs' design refer to programs of National Space Development Agency of Japan (NASDA) and, later, by the Japan Aerospace eXploration Agency (JAXA). Hypersonic Flight Experiment (HYFLEX) was a NASDA unmanned reentry demonstrator which was launched in 1996 from the Tanegashima Space Center by a J-I expendable rocket. It was a successor of OREX and was a precursor for the HOPE-X concept [26]. HYFLEX was a lifting body laboratory to gather data on aerodynamic heating and pressure loads.

3. Present developments

To date there are only two servicing HVs, namely, the X-37 and the Soyuz spacecrafts. As discussed before, the former is a US unmanned reentry spacecraft (winged-body) close to the Space Shuttle, while the latter is the only human-rated capsule operated by Russian Federation [9, 27, 28].

The X-37's aerodynamic design was derived from the Shuttle Orbiter, and hence the X-37 has a similar lift-to-drag ratio (L/D). The X-37 is the smallest and lightest lifting winged vehicle flown to date. It features a forwarded double delta wing and a butterfly tail [9]. The X-37 re-enters Earth's atmosphere and lands automatically. It is the second reusable spacecraft to have such a capability, after the Buran shuttle [14]. The X-37 is now operated by the USAF being transferred to the Defense Advanced Research Projects Agency (DARPA).

The Soyuz spacecraft was designed for the Soviet space program by the Korolev Design Bureau in the 1960s, and it is still in service today. It is currently the only manned space vehicle in the world to support flight to and from the ISS. The spacecraft consists of three parts, namely, orbital module (OM), service module (SM), and reentry module (RM). The OM is a spheroid spacecraft's segment which provides accommodation for the crew during mission. The cylindrical spacecraft's segment is the SM. It features solar panels attached and contains the instruments and engines. Finally, the RM is a small capsule which returns the crew to Earth [27, 28].

4. Future developments

As future developments in HVs, there are Dream Chaser and Space RIDER (both lifting bodies) concepts. They are being developed by the USA and Europe, respectively [13, 29, 30].

The Dream Chaser is a reusable lifting-body spaceplane that can fly autonomously to resupply the ISS with both pressurized and unpressurized cargos. The vehicle is designed to be launched on expendable rockets, return from space by gliding, and autonomously land on conventional runways. The potential further development of the spaceplane includes a human-rated version which would be capable of carrying up from two to seven people to and from LEO. Dream Chaser design is derived from NASA HL-20 lifting body which was itself like the Soviet BOR-4 [1, 14, 31].

The experience and data obtained by Europe so far on Hermes, FESTIP, and FLPP programs served as stepping stones toward a vehicle called Space Reusable Integrated Demonstrator for Europe Return (Space RIDER), underdeveloped by ESA [29]. The Italian Space Agency (ASI), with the project being led by the Italian Aerospace Research Centre (CIRA), presented its own Programme for Reusable In-orbit Demonstrator in Europe (PRIDE) to develop the prototype named Space RIDER [30]. It is an unmanned spacecraft aiming to provide the ESA with affordable and routine access to space.

5. Designing hypersonic vehicles and preset book aims

Such a limited number of operating HVs are due to the high operative cost and, especially, to the complexity in designing such vehicles, especially for human-rated missions. Indeed, HVs' design is an extremely challenging process involving several disciplines, e.g., aerodynamic, aerothermodynamic, control, avionics navigation systems, propulsion, and structure. As well known, these disciplines are strongly coupled with one another and generally influence each other because they involve antagonistic objectives. Therefore, it is expected that synergistic interactions, between vehicle subsystems and functions, can produce an optimized multidisciplinary vehicle design with significant performance and economic improvements [32–40].

This suggests using specific methodologies to assess trade-off analyses between the enabling disciplines as the only way to obtain a satisfactory (global optimal) vehicle design, referred to as multidisciplinary design optimization (MDO) [41–48].

The book aims at highlighting that the design of HVs must pass from a conventional design to a more complex and challenging highly integrated design framework, according to the MDO approach.

Several chapters in the present book focused attention on this fundamental topic, especially for what concerns the design of scramjet-propelled vehicle configurations. For instance, the design and optimization about the integration of airframe-propulsion design issue are discussed as well as the design of vehicle TPS with parametric integral soft object-based procedure. Anyway, investigations of more conventional topics are also provided in the book, as numerical simulations of base pressure and drag of typical reentry vehicles.

In this framework, the ambition of the present book is to support industries, research centers, and space agencies in their own design and development of next-generation HVs. Therefore, this book is recommended for both students and research engineers involved in all design phases, typical for hypersonic vehicles.

Author details

Antonio Viviani and Giuseppe Pezzella*
Department of Engineering, University of Campania "L. Vanvitelli", Aversa, Italy

*Address all correspondence to: giuseppe.pezzella@unicampania.it

References

[1] Jenkins DR, Landis T, Miller J. American X-Vehicles: An Inventory—X-1 to X-50. NASA Special Publication; Monographs in Aerospace History. No. 31. SP-2003-4531. Centennial of Flight ed. Washington, DC: NASA History Office; 2003

[2] Hicks JW, Trippensee G. NASA Hypersonic X-Plane Flight Development of Technologies and Capabilities for the 21st Century Access to Space. AGARD Future Aerospace Technology in Service to the Alliance; 14-18 April 1997; Paris, France; 1997

[3] United States General Accounting Office. National Aero-Space Plane: A Technology Development and Demonstration Program to Build The X-30. Report USGAO/NSIAD-88-1221988. pp. 35-40

[4] Ovallos FG, Smits AJ. Challenges for innovation in aerodynamics. In: 14th Australasian Fluid Mechanics Conference; 10-14 December 2001. Adelaide, Australia: Adelaide University; 2001

[5] Lockwood MK. Overview of Conceptual Design of Early Venturestar Configurations. AIAA-2000-1042; 2000

[6] Hollis BR, Thompson RA, Murphy KJ, Nowak RJ, Riley CJ, Wood WA, et al. X-33 aerodynamic computations and comparisons with wind-tunnel data. Journal of Spacecraft and Rockets. 2001;**38**(5):684-691. DOI: 10.2514/2.3753

[7] Miller CG. Development of X-33/X-34 aerothermodynamic data bases: Lessons learned and future enhancements. In: NATO-AVT Symposium on Aerodynamic Design and Optimization of Flight Vehicles in a Concurrent Multi-Disciplinary Environment; Ottawa, Canada. 1999

[8] Nguyen V, Chaudhary AK, Poladian D. X-40A Guidance and Control and Flight Test Results. AAS Paper No. 00-0015; 1999

[9] Chaudhary A, Nguyen V, Tran H, Poladian D, Falangas E. Dynamics and Stability and Control Characteristics of the X-37. AIAA-2001-4383. 2001

[10] Marchin R, Stein J, Muratore J. An overview of the X-38 prototype crew return vehicle development and test program. In: 15th Aerodynamic Decelerator Systems Technology Conference; AIAA-99-1703; 8-11 June 1999; Toulouse, France

[11] Davis MC, White JT. X-43A flight-test-determined aerodynamic force and moment characteristics at Mach 7.0. Journal of Spacecraft and Rockets. 2008;**45**(3):472-484

[12] Lane J. Design Processes and Criteria for the X-51A Flight Vehicle Airframe. RTO-MP-AVT-145. 2007

[13] Howard RD, Krevor ZC, et al. Dream chaser commercial crewed spacecraft overview. In: Proceedings of 17th AIAA International Space Planes and Hypersonic Systems and Technology Conference; San Francisco, CA; 2011

[14] Chertok B. Rockets and People. Vol. I: The NASA History Series. NASA SP-2005-4110. Washington, DC: National Aeronautics and Space Administration NASA History Office of External Relations; 2005

[15] Van Den Abeelen L. Spaceplane HERMES-Europe's Dream of Independent Manned Spaceflight. Switzerland: Springer International Publishing; 2017. ISBN 978-3-319-44472-7

[16] Stadler R. Results for a Fully Reusable TSTO-Launch Vehicle Concept. AIAA-98-1504. 1998

[17] Spies J, et al. ASTRA system concept investigations overview. In: 3rd International Symposium Atmospheric Reentry Vehicle and Systems; 24-018; Les Mureaux, France: AAAF; 2003

[18] Gockel W. RLV demonstrator phoenix. In: 3rd International Symposium Atmospheric Reentry Vehicle and Systems; 24-019; Les Mureaux, France: AAAF; 2003

[19] Kauffmann J, Baiocco P. Future launchers preparatory programme (FLPP)—System concepts for the next generation launcher. In: Space Access International Conference; 21-23 September; Paris, France; 2011

[20] Pezzella G. Aerodynamic and aerothermodynamic design of future launchers preparatory program concepts. Aerospace Science and Technology. 2012;**23**(1):233-249. DOI: 10.1016/J.Ast.2011.07.011

[21] Pezzella G, Marini M, Roncioni P, Kauffmann J, Tomatis C. Preliminary design of vertical takeoff hopper concept of future launchers preparatory program. Journal of Spacecraft and Rockets. 2009;**46**(4):788-799. DOI: 10.2514/1.39193. ISSN 0022-4650

[22] Pezzella G, Marino G, Rufolo G. Aerodynamic database development of the ESA intermediate experimental vehicle. Acta Astronautica. 2014;**94**(1):57-72. DOI: 10.1016/j.actaastro.2013.07.019. ISSN 0094-5765

[23] Baiocco P. Pre-X experimental reentry lifting body: Design of flight test experiments for critical aerothermal phenomena. In: Flight Experiments for Hypersonic Vehicle Development. Educational Notes RTO-EN-AVT-130, Paper 11. Neuilly-Sur-Seine, France: RTO; 2007. pp. 11-1-11-18

[24] JMA L et al. The SHEFEX flight experiment—Pathfinder experiment for a sky based test facility. In: 14th Spaceplane Systems and Technologies Conference. 2006. DOI: 10.2514/6.2006-7921

[25] Russo G, Capuano A. The PRORA-USV program. In: 4th European Symposium of Aerothermodynamics for Space Applications; 15-18 October 2001; Capua, Italy. ESA SP-487. 2002

[26] Yanagihara M, et al. Numerical Analysis of Hypersonic Aerodynamics for Atmospheric Reentry Problems of HOPE And HYFLEX. AIAA Paper 98-0277; 12-15 January 1998; Reno, Nevada; 1998

[27] Ivanov MS. Statistical Simulation of Reentry Capsule Aerodynamics in Hypersonic Near-Continuum Flows. RTO-EN-AVT-194. 2011

[28] Weiland C. Aerodynamic Data of Space Vehicles. Berlin, Heidelberg: Springer-Verlag; 2014. ISBN: 978-3-642-54167-4

[29] Balossino A, Battocchio L, Giacci M, Giudotti G, Rufolo G, Denaro A, et al. Conceptual design of the descent subsystem for the safe atmospheric reentry flight of space rider. In: 7th European Conference for Aeronautics and Space Sciences (EUCASS). 2017

[30] Marini M, Di Clemente M, Guidotti G, Rufolo G, Lambert O, Joiner N, et al. Aeroshape trade-off and aerodynamic analysis of the space-rider vehicle. In: 7th European Conference for Aeronautics and Space Sciences (EUCASS). 2017. DOI: 10.13009/EUCASS2017-416

[31] Ware GM, Cruz CI. Aerodynamic characterizes of the HL-20. Journal of Spacecraft and Rockets. 1993;**30**(5):529-536

[32] Di Giorgio S, Quagliarella D, Pezzella G, Pirozzoli S. An aerothermodynamic design

optimization framework for hypersonic vehicles. Aerospace Science and Technology. 2019;**84**:339-347

[33] Viviani A, Iuspa L, Aprovitola A. Multi-objective optimization for reentry spacecraft conceptual design using a free-form shape generator. Aerospace Science and Technology. 2017;**71**:312-324. DOI: 10.1016/j.ast.2017.09.030

[34] Viviani A, Iuspa L, Aprovitola A. An optimization-based procedure for self-generation of reentry vehicles shape. Aerospace Science and Technology. 2017;**68**:123-134. DOI: 10.1016/j.ast.2017.05.009

[35] Brevault L, Balesdent M, Defoort S. Preliminary Study on Launch Vehicle Design: Applications of Multidisciplinary Design Optimization Methodologies. HAL Id: Hal-01857722. 2017

[36] Dépincé P, Guédas B, Picard J. Multidisciplinary and Multiobjective Optimization: Comparison of Several Methods. HAL Id: Hal-00449605. 2007

[37] Kulfan B, Bussoletti J. "Fundamental" parametric geometry representations for aircraft component shapes. In: Multidisciplinary Analysis Optimization Conferences; September 2006; American Institute of Aeronautics and Astronautics. 2006. DOI: 10.2514/6.2006-6948

[38] Kulfan B. A universal parametric geometry representation method "CST". In: Aerospace Sciences Meetings; January 2007; American Institute of Aeronautics and Astronautics. 2007. DOI: 10.2514/6.2007-62

[39] Kulfan BM. Universal parametric geometry representation method. Journal of Aircraft. 2008;**45**(1):142-158. DOI: 10.2514/1.29958. ISSN 0021-8669

[40] Wang MY, Wang X, Guo D. A level set method for structural topology optimization. Computer Methods in Applied Mechanics and Engineering. 2003;**192**(1):227-246. DOI: 10.1016/S0045-7825(02)00559-5. ISSN 0045-7825

[41] Hansen N. The CMA evolution strategy: A comparing review. In: Lozano JA, Larranaga P, Inza I, Bengoetxea E, editors. Towards A New Evolutionary Computation. Advances on Estimation of Distribution Algorithms. Springer; 2006. pp. 75-102

[42] Sheffer S, Dulikravich G. Constrained optimization of three-dimensional hypersonic vehicle configurations. In: 31st Aerospace Sciences Meeting; January 1993; American Institute of Aeronautics and Astronautics. 1993. DOI: 10.2514/6.1993-39

[43] Landon M, Hall D, Udy J, Perry E. Automatic supersonic/hypersonic aerodynamic shape optimization. In: 12th Applied Aerodynamics Conference, Fluid Dynamics and Colocated Conferences; June 1994; American Institute of Aeronautics and Astronautics. 1994. DOI: 10.2514/6.1994-1898

[44] Ueno A, Suzuki K. CFD-based shape optimization of hypersonic vehicles considering transonic aerodynamic performance. In: 46th AIAA Aerospace Sciences Meeting and Exhibit, Aerospace Sciences Meetings; January 2008; American Institute of Aeronautics and Astronautics. 2008. DOI: 10.2514/6.2008-288

[45] Dominic D, Erwin M. Optimization of entry-vehicle shapes during conceptual design. Acta Astronautica. 2014;**94**(1):198-214

[46] Iuspa L, Scaramuzzino F, Petrenga P. Optimal design of an aircraft engine mount via bit-masking oriented genetic algorithms. Advances in Engineering Software. 2003;**34**:707-720. ISSN: 0965-9978

[47] Iuspa L, Scaramuzzino F. A bit-masking oriented data structure for evolutionary operators implementation in genetic algorithms. Soft Computing. 2002;5:58-68. ISSN: 1432-7643

[48] D'Amato E, Notaro I, Mattei M. Optimal flight paths over essential visibility graphs. In: Luque A, Hegedus S, editors. Handbook of Photovoltaic Science and Engineering. 2nd ed. Chichester: Wiley; 2011. pp. 169-217. DOI: 10.1002/978047974704.ch5

Section 2

Multidisciplinary Design Optimization

Chapter 2

Parametric Integral Soft Objects-based Procedure for Thermal Protection System Modeling of Reusable Launch Vehicle

Andrea Aprovitola, Luigi Iuspa and Antonio Viviani

Abstract

The present paper deals with a modeling procedure of a thermal protection system (TPS) designed for a conceptual reusable launch vehicle (RLV). A novel parametric model based on a scalar field created by a set of soft object primitives is used to assign an almost arbitrary seamless distribution of insulating materials over the vehicle surface. Macroaggregates of soft objects are created using suitable geometric supports allowing a distribution of coating materials using a limited number of parameters. Applications to different conceptual vehicle configurations of an assigned thickness map and materials layout show the flexibility of the model.

Keywords: reusable launch vehicles, reentry aerodynamics, integral soft objects, hypersonic flow, thermal protection system

1. Introduction

Currently a number of projects related to the development of reusable launch vehicles (RLV) both single-stage-to-orbit (SSTO) and two-stage-to-orbit (TSTO) are ongoing. This trend relates to objectives of future space missions that demand to improve vehicle operability, reducing at the same time flight costs of putting payload into orbit. Several preliminary studies/experiments related to this design scenario have been carried on. The European Space Agency developed two demonstrators, the EXPERT (European eXPErimental Re-entry Test-bed) program and the Intermediate eXperimental Vehicle (IXV), which performed an atmospheric lifting reentry from orbital speed [1]. Besides, an unmanned lifting body developed by Boeing X-37B has been put in orbit by an Atlas-5 rocket and performed a successful lifting-guided reentry. Furthermore, a growing demand for space tourism has emerged also in recent years [2]; therefore, a great deal of research effort has been put to design RLV as blended wing bodies also allowing a conventional and more comfortable landing on runways. The main requirements currently considered for RLV design are (i) to perform very low-g (nearly 1.5 g) reentry; (ii) to adopt a lightweight (passive), fully reusable thermal protection system (TPS) to withstand several flights without any replacement; and (iii) to provide vehicle autonomy to land at a predefined location for crew rescue

[3, 4]. In order to fulfill all those requirements, the duration of reentry flight increases and consequently the integrated heat load absorbed by the structure [3].

The above consideration incidentally demands a trade-off among several nonlinear conflicting design objectives, also satisfying a number of constraint functions. As an example, the design of the TPS of an RLV performing a suborbital lifting reentry requires a mandatory compromise between the maximum allowed peak heating and the integrated heat load. This requirement may conflict with the adoption of a fully reusable TPS, either limiting the choice of material category or penalizing the total mass. In preliminary design practice, thousands of design configurations are typically evaluated by an optimization algorithm to find the best fit [5–11]. Therefore, a preliminary appraisal of vehicle performances is commonly performed using high-efficiency, low-order fidelity methods that give a support to a multidisciplinary analysis performed with a computational effort which fit the typical timeline of the conceptual design phase [11]. In current studies, TPS sizing is performed using several simplified assumptions, carrying out a one-dimensional heat conduction analysis with panel thickness modeled using stackups of different materials [12].

The aerothermal environment is a basic design criterion for either TPS sizing or choice of materials [13, 14]. Several works dealing with TPS sizing have been published in literature. Lobbia [8] determined the sizing of a TPS in the framework of a multidisciplinary optimization. Material densities and maximum reuse temperature were computed. TPS mass was estimated assuming the category of materials used for the space shuttle and thickness distribution assigned on a review of HL-20 materials for each component. Trajectory-based TPS sizing has been proposed by Olynick [13] for a winged vehicle concept. The heating peak was determined considering an X-33 trajectory, discretized in a number of fixed waypoints. The resulting aerothermal database was used as an input for a one-dimensional conduction analysis, and several one-dimensional stackups of different materials representative of TPS were consequently sized. Bradford et al. [14] developed an engineering software tool for aero-heating analysis and TPS sizing. The tool is applicable in the conceptual design phase for reusable, non-ablative TPS. The thermal model was based on a one-dimensional analysis, and TPS was modeled considering a stackup of ten different material layers. Mazzaracchio [15] proposed a method to perform the sizing of a TPS depending on the locations of ablative and reusable zone on a TPS considering the coupling between trajectory and heat shield. Multidisciplinary analysis, integrating a procedural NURBS-based shape representation, is adopted for a preliminary design [3]. NURBS parameterization allows a simple control over the aerodynamic shape using a limited number of sensitive design parameters acting as geometrical modifiers.

However, derivation of a unique parameterization to describe the overall changes of geometry resulting from a shape optimization is not always possible, and several surfaces are used to parameterize different parts of the geometry. Implicit surfaces are a powerful and alternative tool for creating shapes due to their smooth blending properties enabling creation of arbitrary shape. In the present work, a soft object-derived representation for TPS thickness and material attribution is introduced. According to the legacy formulation of this technique, originally developed in computer graphics for the rendering of complex organic shapes [16], three-dimensional object surfaces are (implicitly) obtained by defining a set of source points (or even more complex varieties) irradiating a potential field that is subsequently tracked according to an assigned isosurface. Following a quite different paradigm developed in [17], the full potential field irradiated by a set of by-dimensional soft objects is congruently mapped on a discretized RLV shape. The methodology is able to create arbitrary TPS distributions seamlessly increasing the thickness where critical heat loads are experienced and dropping out elsewhere. A similar, slightly modified procedure is also applied to create an arbitrary binary map of different TPS materials that may be operated independently (or synchronized) with the thickness distribution. The present formulation is formalized in the framework of a parametric model which exploits simple variations of parameters to perform the soft object mapping over discretized surface. Applications of the developed procedure to different arbitrary vehicle shape show the flexibility of the method.

2. Soft objects definition

Soft objects constitute a modeling technique which typically represents a domain using a scalar field, namely, a field function F, defined over a three-dimensional space. An implicit surface S defined as

$$S = \{\boldsymbol{x} \in \mathbb{R}^3 | F(\boldsymbol{x}) = T\} \tag{1}$$

that is, an isosurface S of the field function F specified by the threshold T represents an object instance using a raster conversion algorithm. Soft object modeling overcomes the drawback given by the parametric surfaces; that is, they automatically allow a self-blending between different primitives. Therefore, complex shapes can be modeled defining $n \geq 1$ potential field f_i, with origin in points $\mathbf{x}_i$, and the blending among them is formally accounted by the algebraic summation of their potential fields f_i [18]:

$$F(d) = \sum_{i=1}^{n} f_i(d_i) \tag{2}$$

A commonly adopted notation

$$F_i(d) = f_i \circ d_i \tag{3}$$

composes the distance metric d_i (which determines the shape of the objects associated to the key point $\mathbf{x}_i$), with the field function f_i, being $\mathbf{x}$ the point of space in which the function is evaluated:

$$d_i = \frac{||\boldsymbol{x} - \boldsymbol{x}_i||_k}{r_i} \tag{4}$$

A more powerful representation used in soft object modeling is based on morphological skeleton that synthesizes the morphological properties of a given domain. A skeleton S_k can be defined as a basic geometric entity (such as points, segments, and plain closed domains) around which more complex shapes can be created once the distance function is provided. The simplest soft object was introduced by Blinn that originally proposed the "blobby molecule," an isotropically decaying Gaussian function modulated in strength and radius [16]:

$$f(d_i) = exp\left(-\frac{d_i^2}{2}\right) \tag{5}$$

where d is the Euclidean distance (k = 2 in Eq. (4)). Blobby molecule is a soft object defined around a point skeleton, and its field function has an infinite support.

This aspect affects the computational effort in a practical implementation, because it has to be evaluated in all points of the space. However, in literature, several finite support potential functions have been proposed for different modeling purposes. Wyvill et al. [19] developed the following field function:

$$f(d) = \begin{cases} 1 - \frac{22}{9}d^2 + \frac{17}{9}d^4 - \frac{4}{9}d^6 & d^2 < 1 \\ 0 & otherwise \end{cases} \tag{6}$$

Blanc [18] proposed another field function introducing an internal hardness factor p, which tunes the blending between two different blobs. A higher value of p makes a blob stiffer in the blending, while a low hardness factor generates larger rounded shapes [17]:

$$f(d) = \begin{cases} 1 - \frac{9d^4}{p + (9/2 - 4p)d^2} & d^2 \leq 1/4 \\ \frac{(1-d^2)^2}{3/4 - p + (3/2 + 4p)d^2} & 1/4 < d^2 \leq 1 \end{cases} \tag{7}$$

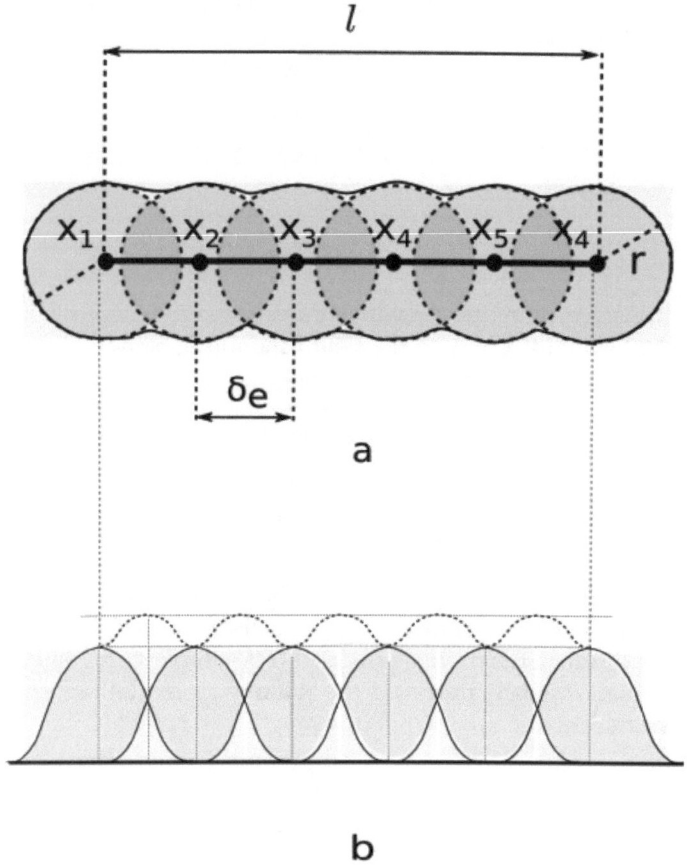

Figure 1.
Support (a) and strength field (b) of a stick created by a superposition of n = 6 point source blobs.

The field function f_i used in the present work has a finite support and assumes normalized values in the range between 0 and 1 [18]:

$$f(d) = \begin{cases} \frac{1}{2} + \frac{1}{2}\frac{arctan(p - 2pd)}{arctanp} & d < 1 \\ 0 & d \geq 1 \end{cases} \quad (8)$$

2.1 Two-dimensional integral soft object for TPS modeling

Two-dimensional soft objects preserve self-blending property. **Figure 1a, b** shows the support and the strength field, respectively, created superposing n = 6 discrete point source blobs with radius r, with origins in key point x_i. If $\delta e < 2r$, two or more blobs superposes, and the strength of the potential field is obtained summing up the strengths of each blob (see **Figure 1b**). A set of n blobs represents a too complex entity if used to model a parametric variation of shape (a single blob is characterized by five independent parameters, i.e., scalar coordinates of centers, strength, and radius). Therefore, blobs can be conveniently and easily arranged in macroaggregates with key points placed on a geometric segment (straight or curved) denoted from now on as "sticks." The point source blobs emulates a segment skeleton with the distance function expressed by Eq. (4) (see **Figure 1a**). However, a simple algebraic summation of potential fields creates a stick support having "bulges." Increasing the number of blobs, the shape of the support becomes

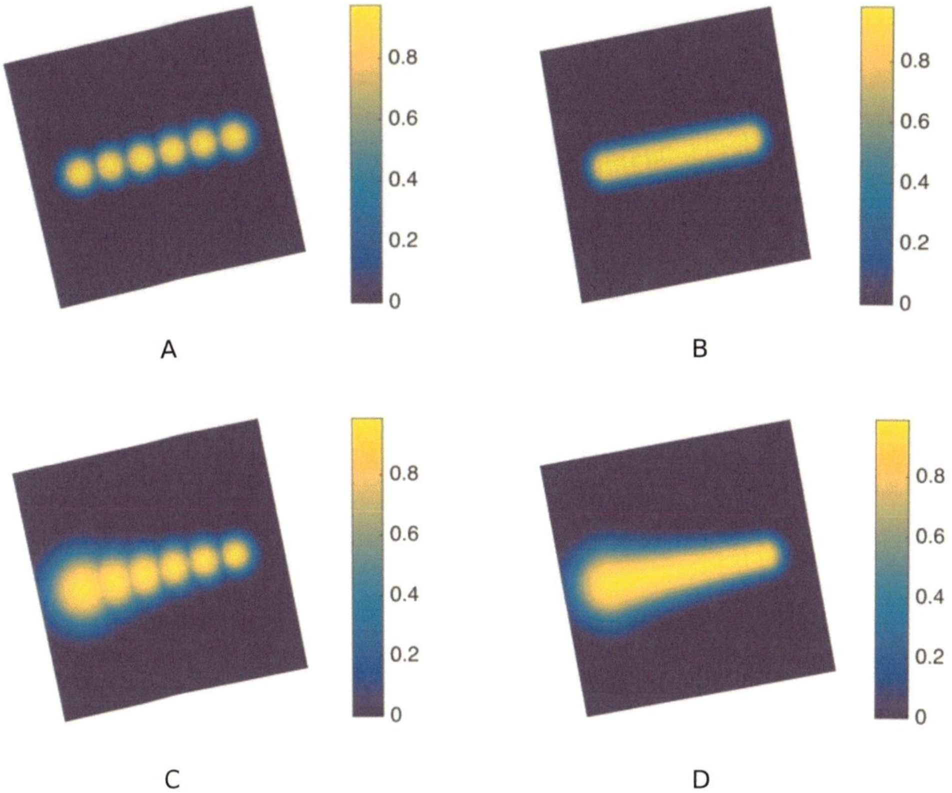

Figure 2.
Stick primitives obtained with nblob = 6 and 20: constant radius (a, b); variable radius (c, d). The stick support becomes more regular increasing nblob; the strength field remains bounded to unit value.

more regular, but the strength of the field function diverges. The above drawback is overcome modifying the definition of potential field given by Eq. (2) with the relation:

$$F_j(P) = \max_{\forall P} \left(F_{j-1}(P), G_j(P)\right) j = 1, \cdots, n_{blobs} \tag{9}$$

Equation (9) where F0(P) = 0 expresses the global potential field $F_j(P)$ irradiated by a set of j blobs at a generic point P of space placed at a distance d from the key points, as the max between the previous j − 1 potentials accounted by the assembly layer $F_{j-1}(P)$ and the current potential G_j over the plane disk of radius r:

$$G_j(P) = \begin{cases} f(P) & d < r \\ 0 & otherwise \end{cases} \tag{10}$$

Figure 2a, **b** shows the support and the strength field of a two-dimensional stick primitive obtained with n_{blob} = 6 and 20, respectively, computed with Eq. (8). By increasing the number of blob on a stick, the strength of F is still bounded to a maximum unit value. **Figure 2c**, **d** shows the same behavior for a tapered primitive having a linear variation of the blob radius along the axis of stick. Therefore, a seamlessly blending of blobs, with a bounded strength, is obtained adopting Eq. (9). The procedure proposed here relies on a similar idea to the one developed in [17] to generate self-stiffened structural panels. Specifically, rather than modeling an object tracking an iso-contour of its potential field, the full integral field generated by a set of blobs spatially arranged on a two-dimensional grid generates a smoothly varying field.

3. RLV shape modeling

A generic shape of an RLV is represented by a grid formed by a quadrangular and/or by either degenerated triangular panel grid. Grid points are obtained using a proprietary procedure that authors fully detailed in [20, 21]. Without going into details of the shape model, we remark that the mesh arrangement over the RLV surface is obtained with no NURBS support surface: a three-dimensional parametric wireframe is created using cubic rational B-splines [22] and used to reconstruct computational surface grid. The control parameter allows a wide range of shape variations to handle different design objectives (thermal or dynamical) for a reentry mission. Grid topology is equivalent to a spherical surface with no singularities (open poles) and allows a mapping of the points in UV coordinates over an equivalent cylindrical surface. The above considerations ensure a topologically invariant shape.

4. Soft object design of TPS

4.1 Rationale

The modeling procedure for the TPS is defined starting from the definition of a set of soft objects which are represented on the topological map associated with the current morphology of the object, as shown in **Figure 3**. Consequently, the supports of the sticks are adjusted according to the normalized dimensions relative to this map. The topological map is emulated introducing a two-dimensional grid (from

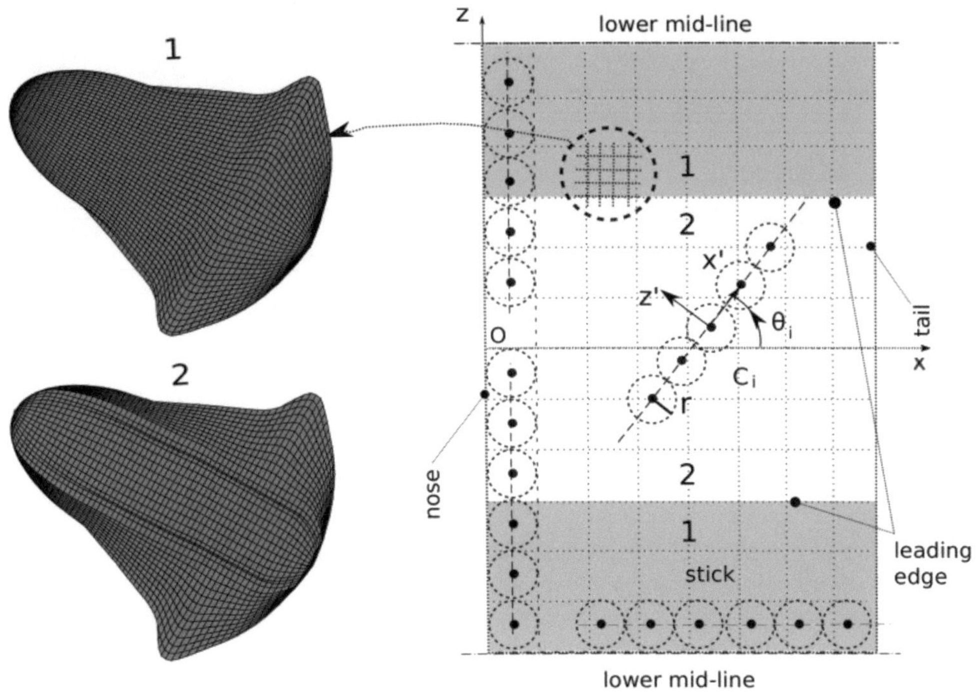

Figure 3.
Morphological (left) vs. topological map (right).

now, denoted as B-grid) having the same topology tree than the vehicle open grid (number of points, panels, and connectivity) but unit size. A geometric mapping between the B-grid and the vehicle grid is established, and elements of B-grid are univocally mapped onto corresponding elements of the vehicle surface (see **Figure 3**). Therefore, each centroid of panels which belong to topological map has the same neighboring points either on the topological or morphological map. Several stick primitives are emulated on B-grid placing a number of n equally spaced isotropic blobs, with radius r and length l, respectively, in a normalized unit. Stick emulation is performed by overlapping n blobs using the special formulation reported in [17] that ensures a convergent envelope of the finite support and a limited value of the blob strength. An exemplificative spatial distribution of sticks on the B-grid is shown in **Figure 3**.

Position and orientation of each stick are determined by assigning coordinates of centers C_i and precession angles θ_i, respectively, with respect to a Cartesian frame of reference Oxz oriented as in **Figure 3**. Therefore, a generic distribution of sticks created on vehicle grid is equally mapped on the vehicle surface whatever is the morphological map considered. In the present case, gray-colored regions (1) denote points of the B-grid mapped on the windward side of RLV shape (see **Figure 3**), while white regions (2) relate to leeward regions of the vehicle. Regions of vehicle surface mainly subjected to heating peaks during the reentry maneuver are (i) nose, (ii) leading edge, and (iii) tail. The global potential field generated by the sticks onto the B-grid is adjusted in a suitable dimensional scale and subsequently mapped on the mesh panels of the vehicle surface grid to obtain an easy and powerful control of the thickness distribution. The proposed methodology is able to create virtually arbitrary TPS distributions and can be easily tuned up to locally increase the thickness where critical heat loads are expected and dropping out elsewhere. A similar, slightly modified procedure is also applied to create an arbitrary binary map

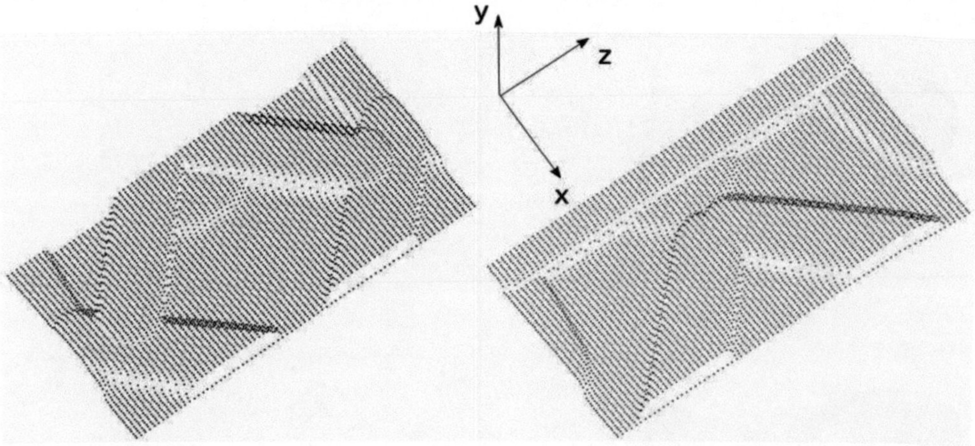

Figure 4.
Arbitrary stick distribution created over the topological map.

distribution of different TPS materials that may be operated independently of the thickness distribution. **Figure 4** shows an arbitrary distribution of stick primitives (not suitable for application purposes) created over the topological map.

The resulting potential field created by the superposition of sticks modulates y-coordinate of grid points as shown in **Figure 4**.

5. Parametric model of thermal protection system

5.1 Thickness modeling

As demonstrative example, a parametric representation of TPS is obtained using a limited set of sticks primitive (n_{stick} = 5), oriented as shown in **Figure 5**. Skin sticks characterized by a large radius and limited strength are spread over the skin surface in longitudinal direction in order to provide a thickness graded baseline. A constant minimum thickness is superposed in all remaining points of B-grid, ensuring a non-zero value in any point of the grid. Furthermore, additional parametric sticks, specifically positioned and oriented to affect thickness in critical regions as nose, leading edge, and trailing edge, complete the support for TPS and create a rational distribution of insulating material suitable with a reentry mission. Parametric position of sticks and axis of orientation are defined by assigning centroid coordinates xc,zc and angle θ_{th}, measured with respect to the system of reference reported in **Figure 5**. Length (l) and strength (th) are expressed with the parametric relations

$$\begin{cases} x_{c,\{q=1,2,3,4,5\}}=\{0.0,0.0,0.0,1.0,1.0\} \\ z_{c,\{q=1,\dots,5\}} = d_{q_{min}} + st_q \bullet \left(d_{q_{max}} - d_{q_{min}}\right) \\ l_{\{q=1,\dots,5\}} = lt_q \bullet d_{q_{max}} \\ th_1 = th'_{min} + pt_1 \bullet \left(th'_{max} - th'_{min}\right) \\ th_{\{q=2,\dots,5\}} = th''_{min} + pt_q \bullet \left(th''_{max} - th''_{min}\right) \end{cases} \tag{11}$$

Skin (q = 1, 2) and nose sticks (q = 3) have a tapered support obtained imposing a linear variation of point source blob radius. Conversely, a constant radius is adopted for the leading edge (q = 4) and trailing edge (q = 5) sticks.

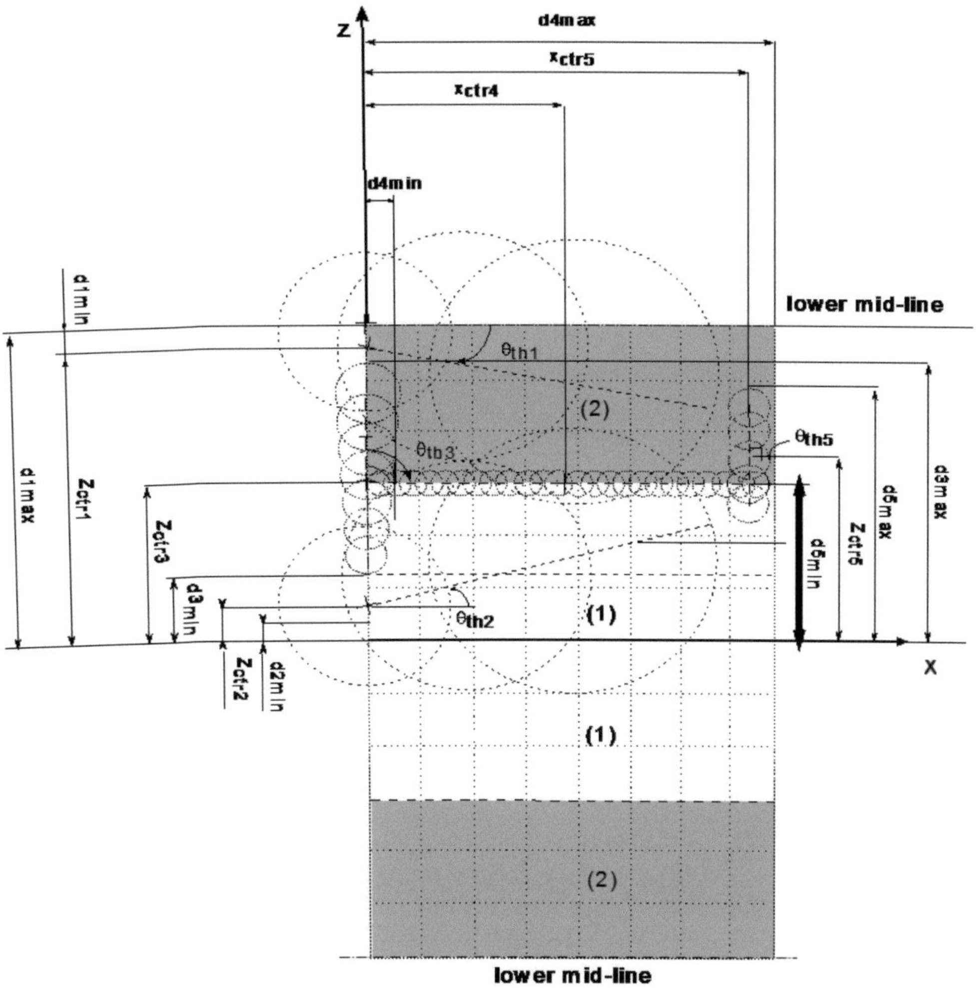

Figure 5.
Arbitrary stick distribution with a longitudinal gradient onto B-grid adopted for TPS modeling.

5.2 Material modeling

A similar but completely independent stick-based parameterization has been also defined to model a dynamic distribution map of different insulating materials, denoted here generically as material 1 and material 0 represented with red and blue colors, respectively. We assume that material 1 outperforms material 0. Therefore, material 1 is adopted on the nose, leading edge, and trailing edge, respectively. Differently than sticks used for thickness distribution, this additional set of primitives returns just binary values used to define specific materials. In this case the field function mth (see relation (12)) assumes a constant value equal to one inside the finite support of a stick and zero elsewhere. The parametric equations which describe material assignments are

$$\begin{cases} mx_{c,\{q=1,2,3,4,5\}=\{0.0,0.0,0.0,1.0,1.0\}} \\ mz_{c,\{q=1,\dots,5\}} = d_{q_{min}} + mt_q \bullet \left(d_{q_{max}} - d_{q_{min}}\right) \\ ml_{\{q=1,\dots 5\}} = mlt_q \bullet d_{q_{max}} \\ mth_{\{q=1,\dots,5\}} = 1 \end{cases} \tag{12}$$

Parameter	Value	Parameter	Value
st_1, ad	0	mt_1, ad	1
st_2, ad	0.01	mt_2, ad	0.01
st_3, ad	0.05	mt_3, ad	0.05
st_4, ad	1	mt_4, ad	1
st_5, ad	0.8	mt_5, ad	0.8
lt_1, ad	1	mlt_1, ad	1
lt_2, ad	0.1	mlt_2, ad	0.1
lt_3, ad	1	mlt_3, ad	1
lt_4, ad	1	mlt_4, ad	1.2
lt_5, ad	1	mlt_5, ad	1
pt_1, ad	1	–	–
pt_2, ad	0.2	–	–
pt_3, ad	0.5	–	–
pt_4, ad	0.2	–	–
pt_5, ad	0.6	–	–
d_{1min}, ad	0.5	d_{1max}, ad	1
d_{2inin}, ad	0.01	d_{2max}, ad	0.3
d_{3min}, ad	0.09	d_{3max}, ad	1
d_{4min}, ad	0.1	d_{4max}, ad	0.5
d_{5min}, ad	0.02	d_{5max}, ad	0.5
th'_{min}, ad	0.07	th'_{max}, ad	0.12
th''_{min}, ad	0.132	th''_{max}, ad	0.25

Table 1.
Parameters adopted in the modeling of TPS configurations of ***Figures*** *7 and 8.*

with normalized parameters reported in **Table 1**.

6. TPS modeling capabilities

The previously introduced modeling procedure has been applied on a conceptual RLV shape created with the model described in Section 4 and detailed in [20, 21]. **Figure 6** shows a topological map obtained for an arbitrarily chosen distribution of stick primitives.

A local thickness is assigned on the nose, the leading edge, and the trailing edge. The topological map shown in **Figure 6** creates a morphologically adaptive TPS on two RLV shapes with different dimensions: (RLV-1) with length ltot = 9.8 m, wingspan ws = 5.6 m, cabin height h = 1.6 m, and (RLV-2) with length ltot = 15 m, wingspan ws = 9.2 m, and cabin height h = 2 m. The parameters characterizing the distribution of thickness and of the materials are reported in **Table 1**. **Figure 7a, b** shows the application of TPS modeling over the first configuration (RLV-1), on leeward (a) and windward (b) surface, respectively. Different colors denote different values of thickness and are represented in a dimensional scale.

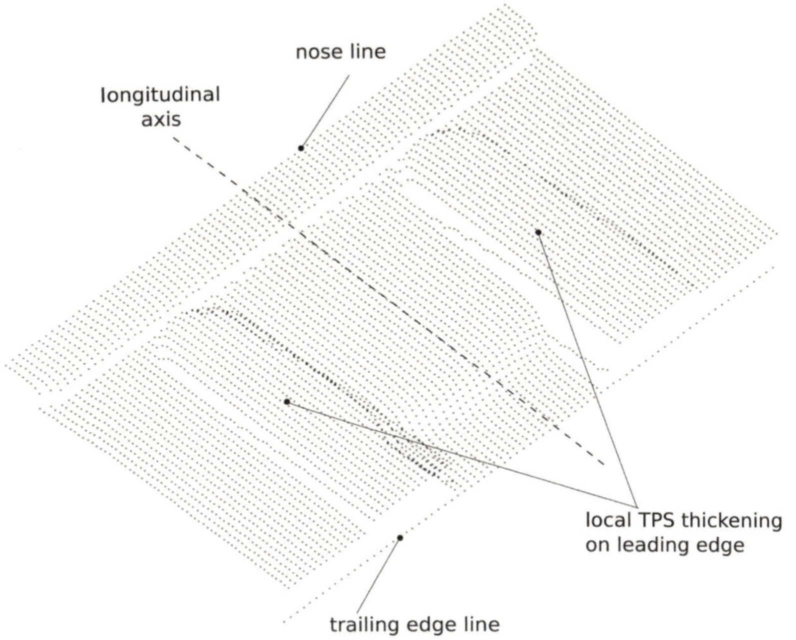

Figure 6.
Topological map created to represent TPS thickness on different RLV configurations.

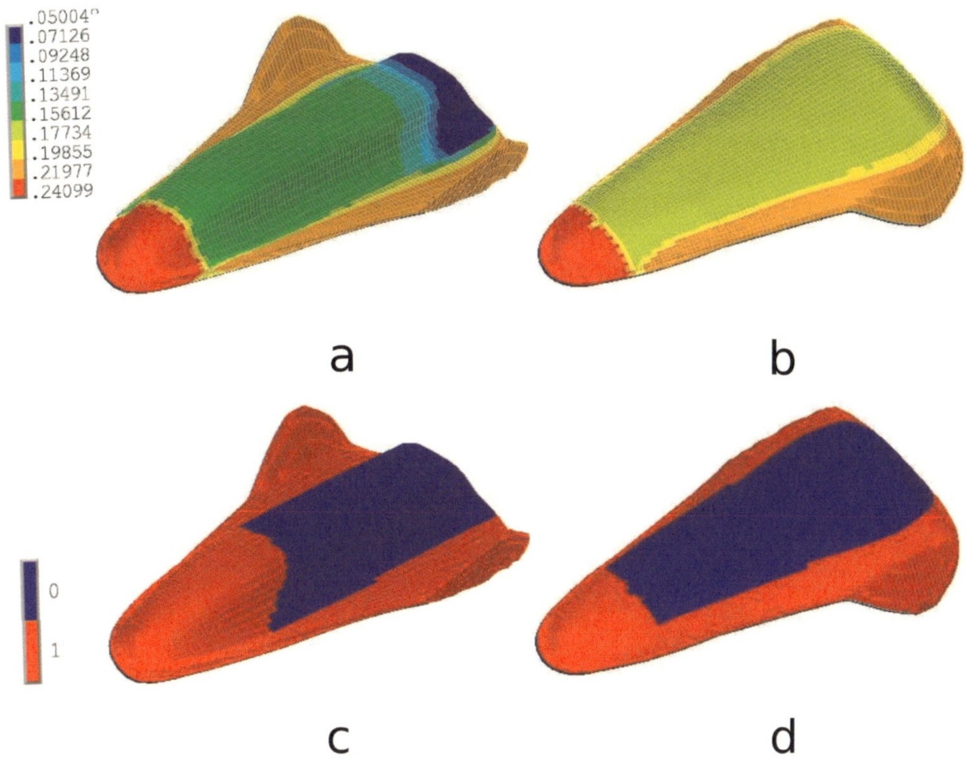

Figure 7.
Example of thickness and material distribution over RLV configuration (RLV-1): (a, b) thickness modulation [m]; (c, d) two material map (red/blue color indicates material 1/0, respectively).

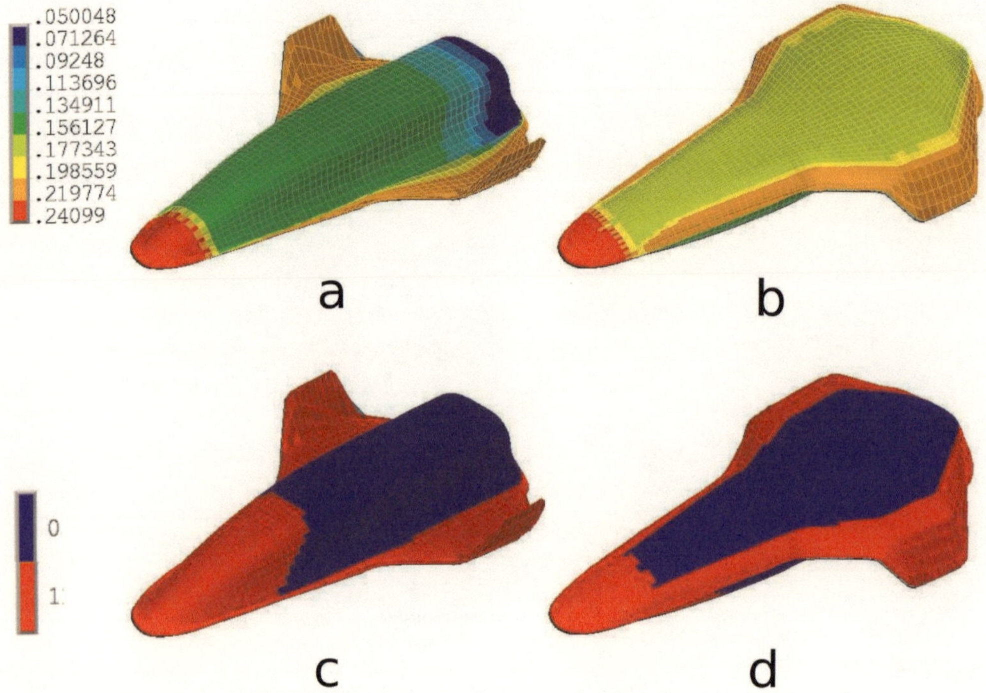

Figure 8.
Example of thickness and material distribution over RLV configuration (RLV-2): (a, b) thickness modulation [m]; (c, d) two material map (red/blue color indicates material 1/0, respectively).

It can be observed that the thickness map can be easily tuned up for best covering of regions where maximum heat loads occur (i.e., the nose and leading edge). **Figure** 7 shows the capability to create arbitrary seamless thickness distribution up to the value of the baseline thickness which has been arbitrarily set equal to th_{min} = 0.05 m (denoted in blue color). This corresponds to a region of the leeward surface not covered by the skin stick. **Figure 7c**, **d** shows the map of two different insulating materials created with Eq. (7). Red colors indicate material 1, which is placed on regions of the vehicle subjected to higher heat loads. Comparisons between **Figure 7a**, **b** and **Figure 7c**, **d** also exhibit the capability of the model to handle independently both the thickness and material distribution. Finally, **Figure 8a**, **b** and **Figure 8c**, **d** show the same blob distribution adopted either for thickness or material modeling applied on a different RLV configuration (RLV-2). The procedure creates, as it was expected, the same TPS distribution both for thickness or materials on two different shapes and is completely independent by their morphology.

7. Conclusions

In the present paper, a special modeling procedure of the TPS designed for a conceptual RLV has been developed. A set of macroaggregates of point source blobs organized in envelopes of finite supports, and with a bounded strength, has been successfully created on the topological map associated with the computational grid. Applications of the modeling procedure to different design configurations highlighted the sensitivity and powerful control to radically change the TPS using a limited number of parameters. The promising capabilities of the developed modeling procedure suggest that the present methodology can give support to a multidisciplinary analysis optionally included in a conceptual design framework. Further developments of the considered procedure are about to be integrated in a companion paper by the authors [23].

Acknowledgements

This work was supported by the Universitá della Campania: "Luigi Vanvitelli."

Conflict of interest

The authors declare that there is no conflict of interest regarding the publication of this chapter.

Author details

Andrea Aprovitola, Luigi Iuspa and Antonio Viviani*
Department of Engineering, Università degli Studi della Campania "L. Vanvitelli", Aversa, Italy

*Address all correspondence to: antonio.viviani@unicampania.it

References

[1] Sziroczak D, Smith H. A review of design issues specific to hypersonic flight vehicles. Progress in Aerospace Sciences. 2016;**84**:1-28. DOI: 10.1016/j.paerosci.2016.04.001

[2] DePasquale D, Charania AC, Olds JR. The emerging orbital space tourism industry: New insight into demand and prospect for success. Space 2006. AIAA Paper No. 2006-7478

[3] Hirschel E, Weiland C. Selected Aero-Thermodynamic Design Problem of Hypersonic Flight Vehicles. Springer-Verlag/AIAA; 2009. DOI: 10.1007/978-3-540-89974-7

[4] Dirkx D, Mooij E. Conceptual Shape Optimization of Entry Vehicles Applied to Capsules and Winged Fuselage Vehicles, Springer Aerospace Technology. Switzerland: Springer International Publishing; 2017. DOI: 10.1007/978-3-319-46055-0

[5] Dirkx D, Mooij E. Optimization of entry-vehicle shapes during conceptual design. Acta Astronautica. 2014;**94**(1): 198-214. DOI: 10.1016/j.actaastro.2013.08.006

[6] Ridolfi G, Mooij E, Dirkx D, Corpino S. Robust multi-disciplinary optimization of unmanned entry capsules. 2012. AIAA Paper No. 2012-5006

[7] Rowell RF, Braun RD. Multidisciplinary conceptual design optimization of space transportation systems. Journal of Aircraft. 1999;**35**(1): 218-226

[8] Lobbia MA. Multidisciplinary design optimization of waverider-derived crew reentry vehicles. Journal of Spacecraft and Rockets. 2017;**54**(1):233-245. DOI: 10.2514/1.A33253

[9] Di Giorgio S, Quagliarella D, Pezzella G, Pirozzoli S. An aerothermodynamic design optimization framework for hypersonic vehicles. Aerospace Science and Technology. 2019;**84**:339-347

[10] Zhang T-t, Wang Z-g, Huang W, Yan L. Parameterization and optimization of hypersonic-gliding vehicle configurations during conceptual design. Aerospace Science and Technology. 2016;**58**:225-234. DOI: 10.1016/j.ast.2016.08.020

[11] Hinman WS, Johansen CT. Rapid prediction of hypersonic blunt body flows for parametric design studies. Aerospace Science and Technology. 2016;**58**:48-59. DOI: 10.1016/j.ast.2016.08.007

[12] Kumar S, Mahulikar SP. Design of thermal protection system for reusable hypersonic vehicle using inverse approach. Journal of Spacecraft and Rockets. 2017;**54**(2):436-446. DOI: 10.2514/1.A33688

[13] Olynick D. Trajectory-based thermal protection system sizing for an X-33 winged vehicle concept. Journal of Spacecraft and Rockets. 1998;**35**(3): 249-256. DOI: 10.2514/2.338

[14] Bradford JE. Thermal protection system sizing and selection for RLV using the sentry code. 2006. AIAA Paper No. 2006-4605

[15] Mazzaracchio A. Thermal protection system and trajectory optimization for orbital plane change aeroassisted maneuver. Journal of Aerospace Technology and Management. 2013; **5**(1):49-64

[16] Blinn JF. A generalization of algebraic surface drawing. ACM Transactions on Graphics. 1982;**68**(3): 123-134

[17] Iuspa L. Free topology generation of self-stiffened panels using skeleton-based integral soft objects. Computers and Structures. 2015;**158**:184-210. DOI: 10.1016/j.compstruc.2015.06.013

[18] Blanc C, Schlick C. Extended field functions for soft objects. In: Proceedings of the Implicit Surface'95; Grenoble, France; 1995. pp. 21-32

[19] Wyvill G, Pheeters C, Wyvill B. Data structures for soft objects. The Visual Computer. 1986;**2**:227-234

[20] Viviani A, Iuspa L, Aprovitola A. An optimization-based procedure for self-generation of re-entry vehicles shape. Aerospace Science and Technology. 2017;**68**:123-134. DOI: 10.1016/j.ast.2017.05.009

[21] Viviani A, Iuspa L, Aprovitola A. Multi-objective optimization for re-entry spacecraft conceptual design using a free-form shape generator. Aerospace Science and Technology. 2017;**71**:312-324. DOI: 10.1016/j.ast.2017.09.030

[22] Florez H, Borges B. Scalar and Parametric Spline Curves and Surfaces. Rijeka, Croatia: Intech Open; 2018. DOI: 10.5772/intechopen.74929

[23] Aprovitola A, Iuspa L, Viviani A. Thermal protection system design of a reusable launch vehicle using integral soft objects. International Journal of Aerospace Engineering. 2019:(3). DOI: 10.1155/2019/6069528

Chapter 3

Airframe-Propulsion Integration Design and Optimization

Yao Zheng, Shuai Zhang, Tianlai Gu, Meijun Zhu, Lei Fu, Minghui Chen and Shuai Zhou

Abstract

Airframe-propulsion integration design is one of the key technologies of the hypersonic vehicle. With the development of hypersonic vehicle design method, CFD technology, and optimization method, it is possible to improve the conceptual design of airframe-propulsion integration both in accuracy and efficiency. In this chapter, design methods of waverider airframes and propulsion systems, including inlets, nozzles, isolators, and combustors, are reviewed and discussed in the light of CFD analyses. Thereafter, the Busemann inlet, a three-dimensional flow-stream traced nozzle, and a circular combustor together with a cone-derived waverider are chosen to demonstrate the airframe-propulsion integration design. The propulsion system is optimized according to the overall performance, and then the component such as the nozzle is optimized to obtain a better conceptual configuration.

Keywords: hypersonic vehicles, airframe-propulsion integration, conceptual design, optimization, computational fluid dynamics (CFD)

1. Introduction

Airframe-propulsion integration design is one of the key techniques of air-breathing hypersonic vehicles [1] to reduce overall drag and achieve positive thrust margins at hypersonic speeds [2]. The engine and airframe aerodynamics therefore become highly coupled [3]. Airframe-propulsion integration methodologies for the hypersonic vehicle have been extensively studied by many researchers [4–7]. A waverider is any supersonic or hypersonic lifting body that is characterized by an attached, or nearly attached, bow shock wave along its leading edge. Since its high lift-to-drag ratio, the waverider has become one of the most promising designs for air-breathing hypersonic vehicles. In the present study, the cone-derived waverider is used and optimized as the basis for the entire vehicle [8], and the engine is generated maintaining the shock wave attaching to the leading edge.

The design of the scramjet, which is a key part of the hypersonic vehicle technology, involves a lot of subjects. Typically, it includes components such as inlet, isolator, combustor, and nozzle. Considering both good performance of every component and interaction effects between each two components, the design progress becomes quite complicated. To effectively solve these difficulties, the present work proposes a method for integrated design and performance analysis of the scramjet flowpath. Aerodynamic performance and flow fields are analyzed one after another for the scramjet and component.

The whole process is in the order of zero-dimensional thermodynamic analysis [9], quasi-one-dimensional estimated analysis [10], and three-dimensional computational fluid dynamics analysis [11–13]. The scramjet flowpath is designed with an inward-turning inlet [12], a constant-area circular isolator, a circular combustor with a cavity [11], and a three-dimensional flow-stream traced nozzle [13]. Firstly, geometry parameters and flow conditions of both the inlet and the exit for each subsystem are obtained from the result of the stream function analysis and optimization [14]. Secondly, two design codes are developed, one of which is the quasi-one-dimensional estimation program for the combustor and the other is the aerodynamic force and heat estimation for the whole hypersonic vehicle. Lastly, the CFD method is applied for performance analysis of the jaws inlet, back pressure characteristics of the inlet with a constant-area isolator, and flow field characteristics of the combustor with a cavity.

2. Waveriders

In the waverider design, it is the first step to define the generation field and then the streamlines constituting the compression surface of the waverider. In the current study, the design conditions of the vehicle are chosen as follows: height of 25 km and free stream with the inflow Mach number to be 5.0. Thereafter, the shape together with the pressure distribution is determined. Typically, a waverider design process can be divided into: selection and design of the basic flow field in the flow direction, solving of the basic flow field, streamline tracing, and application of the osculating theory in the spanwise direction. After that, points representing streamlines are obtained. Streamlines and compression surface can be generated using CAD tools. For example, in this study, an automatic 3D configuration generation program based on the UG API is developed. Meanwhile, an aerodynamic force estimation program is built. Usually, remodel design of the waverider is needed for a specific purpose.

The basic flow field is usually a steady inviscid supersonic flow one, which is the core of the design of a waverider. Basic flow fields used for waverider design can be

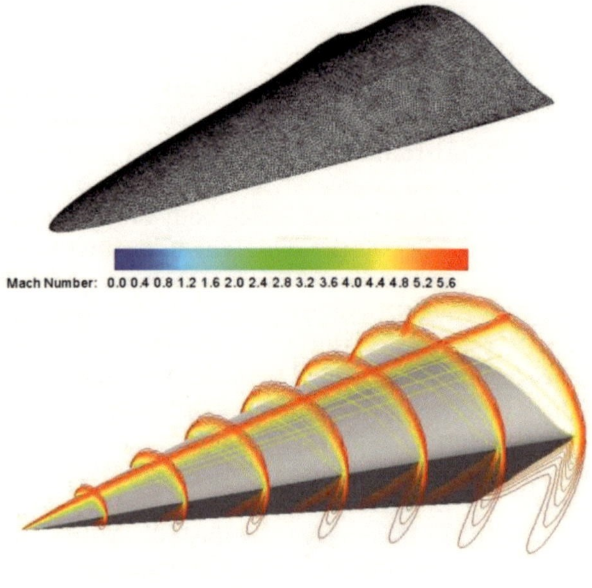

Figure 1.
Cone-derived waverider. (upper) Configuration with surface mesh for rapid estimation. (bottom) CFD simulation under design conditions.

classified into two types: steady two-dimensional (2D) planar or asymmetrical supersonic flow fields and three-dimensional (3D) supersonic flow fields. Since 2D planar or asymmetrical flow fields can be calculated easily by fast calculation methods such as the method of characteristics (MOC), most of the basic flow fields used for rapid design and optimization of waveriders are of the 2D planar or asymmetrical type.

The supersonic flow around a cone at a zero angle of attack, which is known as a conical flow field, is a typical 2D asymmetrical basic flow field. In 1968, Jones et al. [15] first used this kind of flow field to design a waverider known as the cone-derived waverider. The conical flow field is by far the most widely used basic flow field for waveriders, and the cone-derived waverider has become the most widely used waverider owing to the ease of its calculation and better volumetric efficiency than wedge-derived waverider on account of the concave streamlines being closer to the shock wave. **Figure 1** shows our designed cone-derived waverider configuration and numerical simulation results. The surrogate modes for aerodynamic shape optimization were carried out based on the values of lift-to-drag ratio and volumetric efficiency response [8]. In the present study, the commercial software Fluent and Insight are chosen for the numerical simulation and optimization, respectively.

3. Scramjet

Regarding the conceptual design of scramjet, the stream thrust analysis [9] was superior to that of the thermodynamic cycle or first law analyses as it managed to account for several phenomena such as the geometry of the combustor, the velocity, mass of the fuel, and the exhaust outlet pressure not matching the ambient. **Figure 2** briefly introduces the design procedure and method of scramjet.

A scramjet was first designed by using stream thrust analysis, to obtain the overall parameters and flow state parameters at the in-/outlet of each components. When the stream thrust was analyzed, a group of state parameters were determined such as pressure, density, and temperature together with velocity and areas in each component's inlet and outlet. These parameters were delivered to the following two-dimensional components' design of the inlet, isolator, combustor, and nozzle. When the overall dimension of the scramjet internal flow passage was determined, the performance of a scramjet that allowed a supersonic flow to pass through the engine without choking in the inlet throat, combustor, and nozzle were analyzed by the quasi-one-dimensional evaluation program.

Using the above conceptual design method and performance evaluation, the initial design and analysis of a scramjet were performed. The optimization was conducted to generate more practical results based on the specific objectives. The flow chart describing the conceptual design method and optimization process is shown in **Figure 3**. As the design and evaluation of scramjet are highly nonlinear problems,

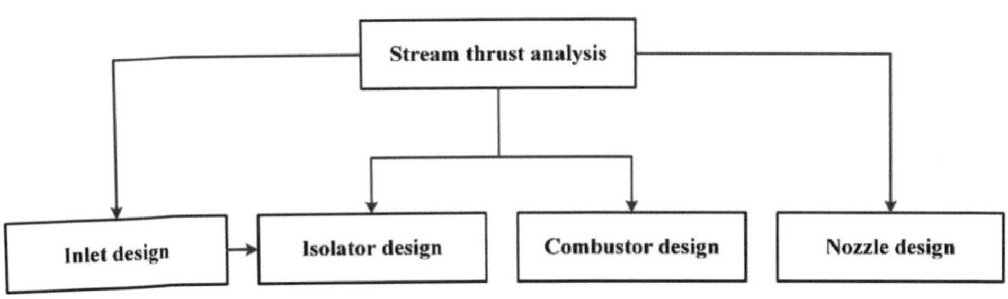

Figure 2.
Flow chart of conceptual scramjet design.

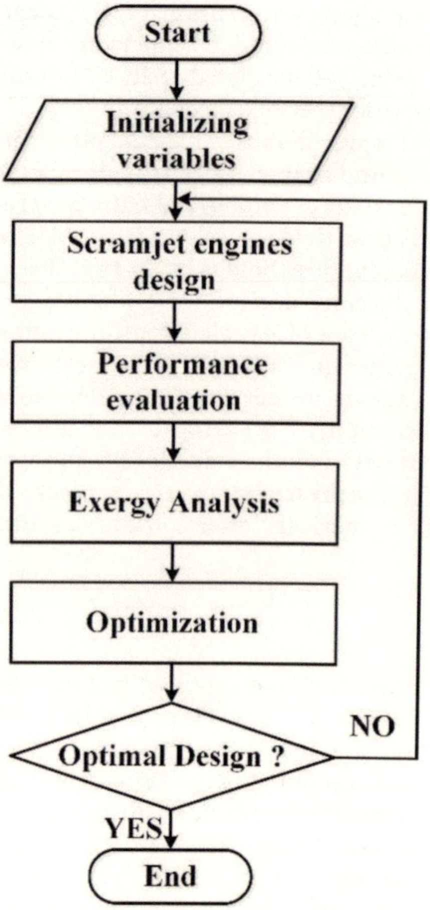

Figure 3.
Flow chart of the conceptual design and optimization process.

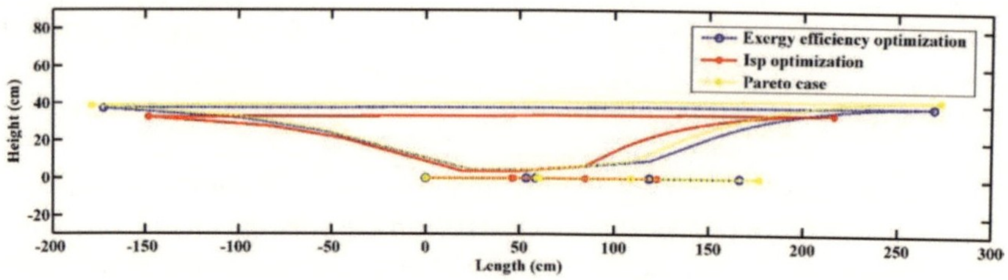

Figure 4.
Comparison of three optimal vehicle geometry shapes.

a multi-island genetic algorithm was chosen as the single-objective optimization algorithm, and a nondominated sorting genetic algorithm was selected as the multi-objective optimization algorithm. After the scramjet was designed and evaluated, the optimization was conducted to study how exergy works in the complex integrated system and to find which design variables play relatively important roles in this evaluation system. **Figure 4** shows the vehicle geometry shapes of different optimization objective cases. Three-dimensional design result of the scramjet can be seen in **Figure 5**. The detailed design methods for each part will be introduced in the following.

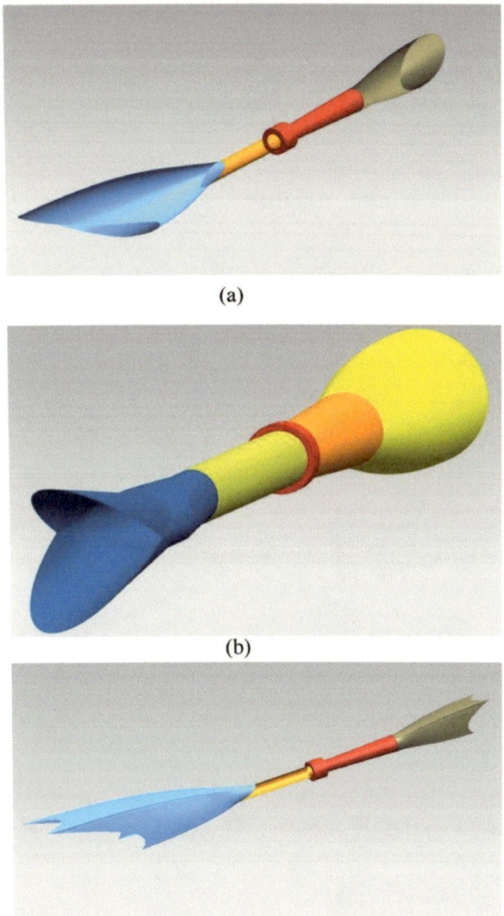

Figure 5.
Conceptual three-dimensional scramjet design: (a) Busemann inlet, circular isolator and combustor, three-dimensional asymmetric nozzle. (b) Jaws inlet, circular isolator and combustor, three-dimensional symmetric nozzle. (c) REST inlet, circular isolator and combustor, three-dimensional asymmetric circle-to-rectangle nozzle.

3.1 Inlets

Various prototypes of hypersonic inlet have been proposed since the 1960s. Kothari introduced the radial deviation parameter and categorized the inlets as inward-turning inlets, outward-turning inlets, and two-dimensional inlets [16]. Unlike the other two groups, inward-turning inlets exhibit accumulation of flows in the central part. The advantages of inward-turning designs, especially those approaching a completely round combustor entrance shape, are several fold. From structural and wetted area perspectives, a more round design provides better performance than a rectangular or two-dimensional configuration. Lower wetted surface area in the combustor for an equivalent level of thrust of course means lower heating loads and lower drag. Moreover, corner flows need be much less of a concern with inward-turning geometries. Low aspect ratios at the isolator, which are characteristic of inward-turning inlets, also result in operational advantages. In **Figure 6**, typical inward-turning inlets such as Busemann, REST, and Jaws inlets were designed and built using CAD tools.

The performance of designed inlets under on and off design conditions was numerically investigated [12]. **Figure 7** shows the comparison on the performance

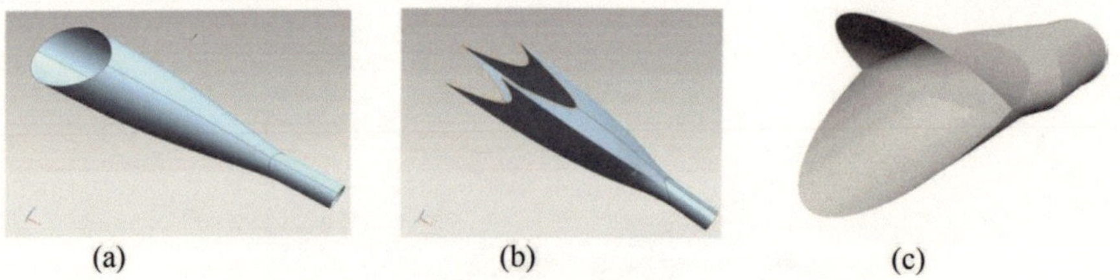

Figure 6.
Several inward-turning inlets design. (a) Busemann. (b) REST. (c) Jaw's.

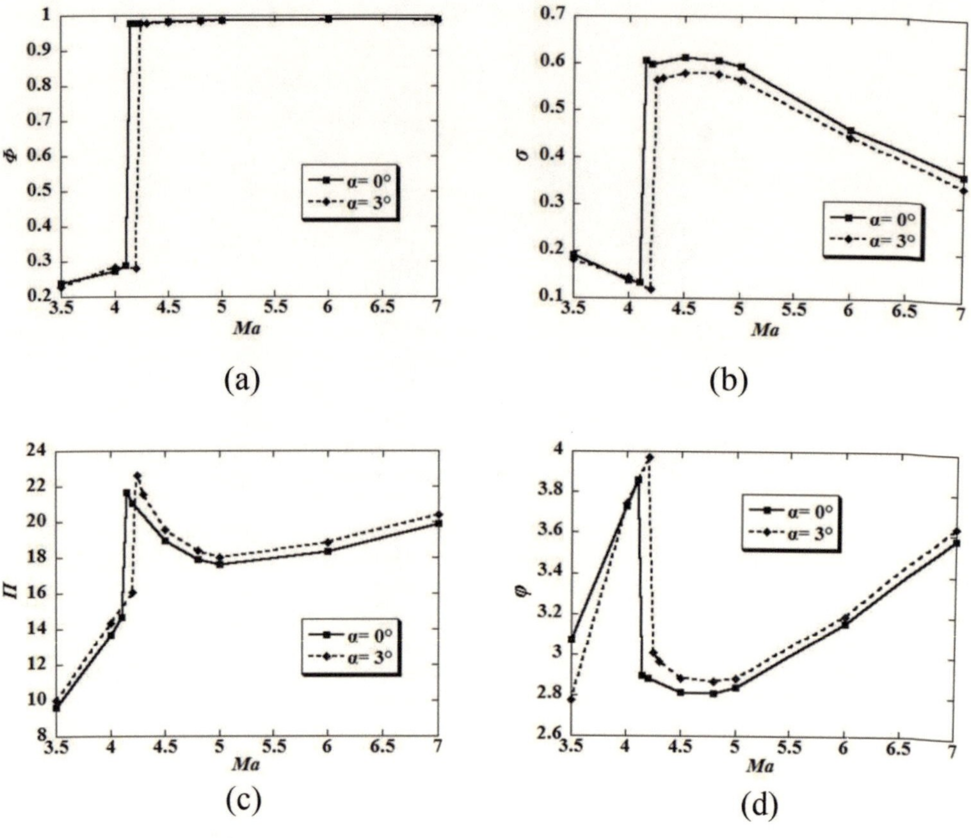

Figure 7.
Performance parameters of Jaws inlet under different inflow Ma: (a) mass capture ratio, (b) total pressure recovery coefficient, and (c) static pressure ratio (d) temperature ratio.

of one Jaws inlet under different inflow Mach numbers and two different angles of attack. In **Figure 8**, numerical simulations were carried out for Jaws inlet under different back pressures. In addition, performances of different inward-turning inlets were also compared using numerical simulations.

3.2 Isolators

An isolator is necessary in scramjet to prevent inlet to unstart under high back pressure due to heat release in the combustor. The backpressure can cause the isolator flow to fluctuate violently. As the back pressure exceeds the critical value, the inlet can unstart, which causes the flow field to become unstable and oscillate unsteadily, with the drag increasing sharply and causing the engine to lose thrust.

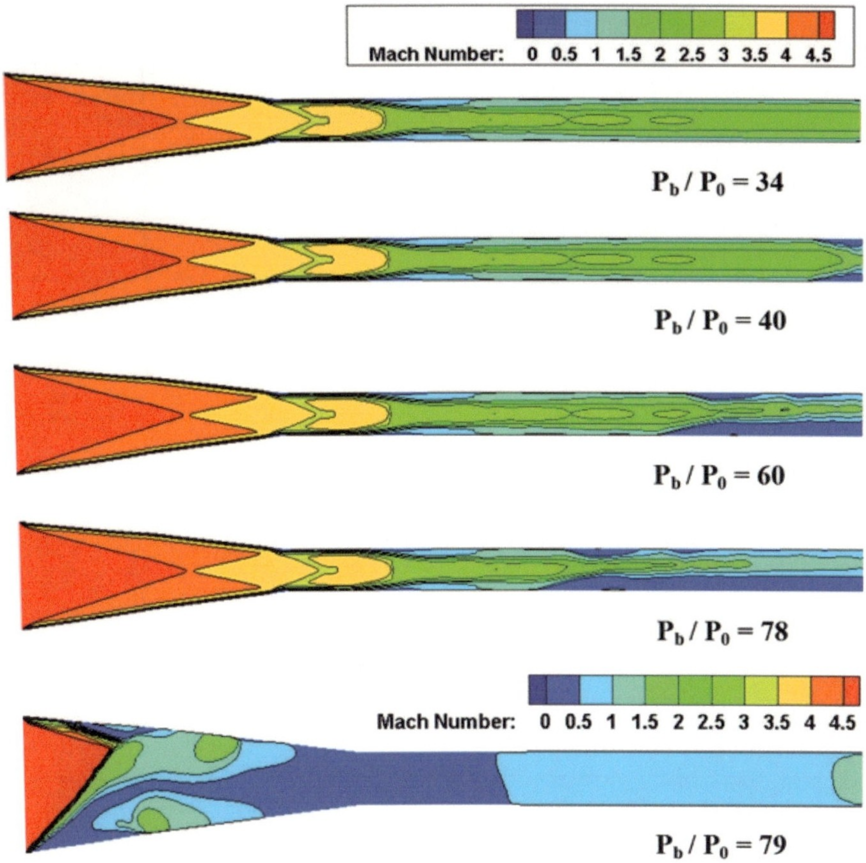

Figure 8.
Mach number contours of Jaws inlet flow fields under different back pressures.

In the isolator, due to the close coupling of the boundary layer and the supersonic core flow through shock waves and expansion waves, the flow structure of a shock train is rather complex even at very simple incoming flow conditions and wall conditions. Correspondingly, it is important to understand the mechanism of the pseudo-shock motion in the isolator. **Figures 9** and **10** show the comparison between numerical simulation results and experimental observation to understand the complex pseudoshock train in the circular and rectangular isolators, respectively. The conceptual design of isolator with a given shape of the inlet is conducted by using the empirical length formula [14]. Meanwhile, the isolator is further truncated with the passive wedge flow control. **Figure 11** shows the simulation results of design without and with the wedge flow control, as shown in the left and right of the figure.

3.3 Combustors

In the combustor, injection fuels mix with incoming air and burn to release large amounts of energy. In the conceptual design, the combustor length includes the ignition length and the combustion length. The ignition length can be obtained by multiplying the ignition delay time which referred to Balakrishnan and Williams [17] by the relative velocity between air and fuel. The combustion length can be modeled based on the study of Hasselbrink [18] and Smith [19]. **Figure 12** shows the conceptual design result of the combustor.

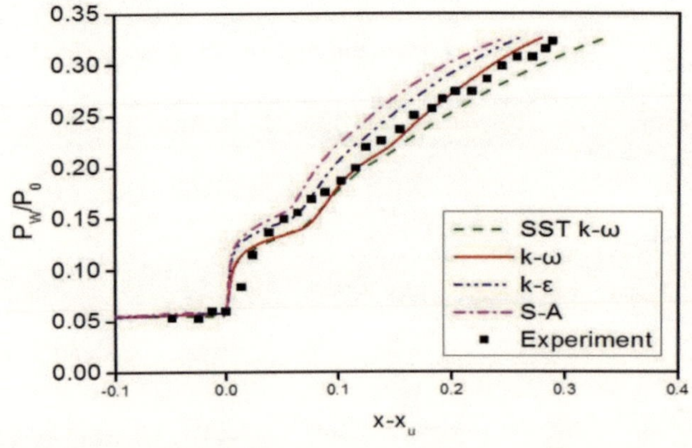

Figure 9.
Pressure distribution along wall using different turbulent models.

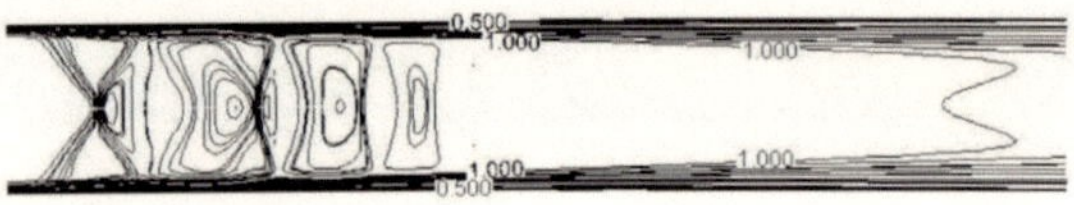

Figure 10.
Contour of Mach number in the central symmetry plane.

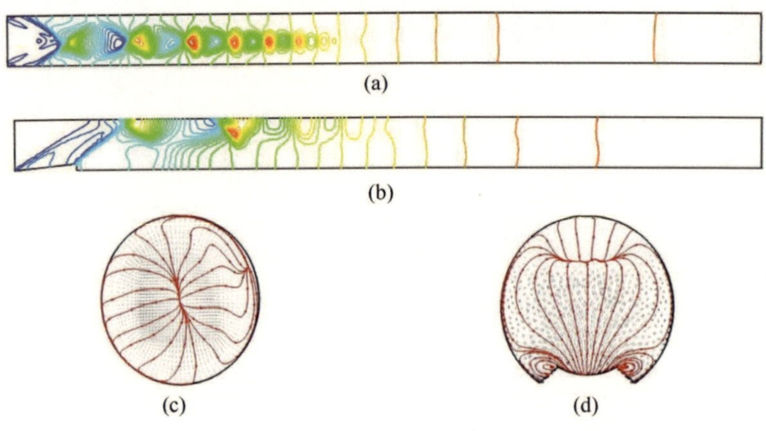

Figure 11.
Simulation result of circular isolator. (a) Pressure contour of baseline design. (b) Pressure contour of baseline design with wedge control. (c) Velocity distribution of baseline design at x = 300 mm. (d) Velocity distribution of baseline design with wedge control at x = 300 mm.

According to the combustion in scramjet engines, the time available for fuel injection, mixing, and combustion is very short. It is important to study the flame holding mechanisms. The presence of normal fuel injector inside the combustor generates a detached normal shock toward the upstream direction of the injector. As a result, there is a formation of separation region which may influence the efficiency of the combustor. As shown in **Figure 13**, numerical simulations were performed to understand the related flow structures. Another alternative method for better-mixing phenomena in scramjet combustor is to use cavity flame holders. Numerical studies on the cavity in the combustor were carried out, as can be seen in **Figure 14**. The fuel injection position was numerically investigated to find out an appropriate value, as shown in **Figure 15**.

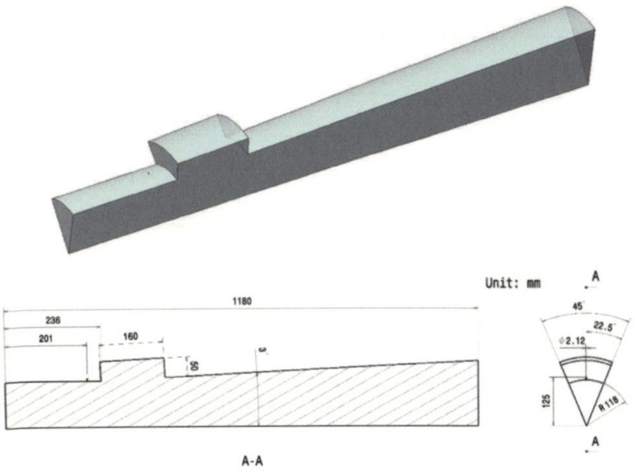

Figure 12.
Geometry of combustor with cavity in 1/8 size.

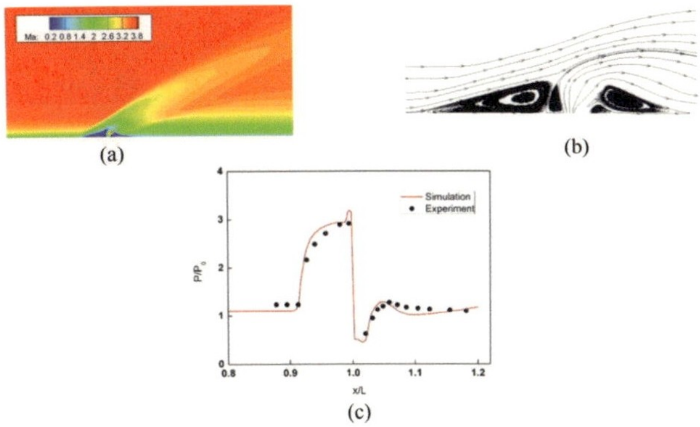

Figure 13.
Simulation of normal inlet under supersonic inflow: (a) Mach contour, (b) streamline, and (c) pressure distribution along wall.

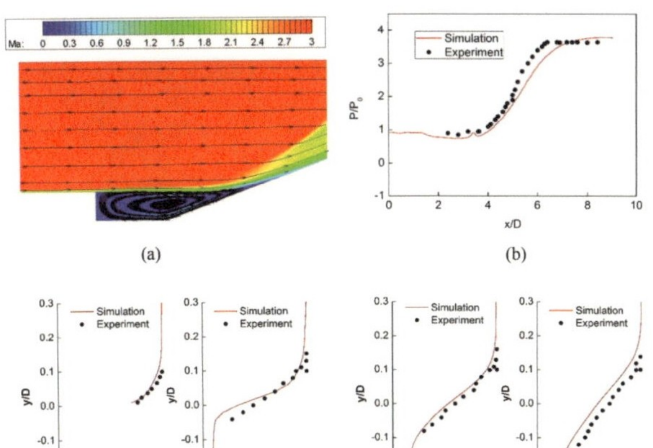

Figure 14.
Numerical simulation results of cavity under supersonic inflow. (a) Ma contour and streamline. (b) Pressure distribution along the cavity wall. (c) Velocity distribution across the shear layer at x = 25.4 and 38.1 mm. (d) Velocity distribution across the shear layer at x = 63.5 and 88.9 mm.

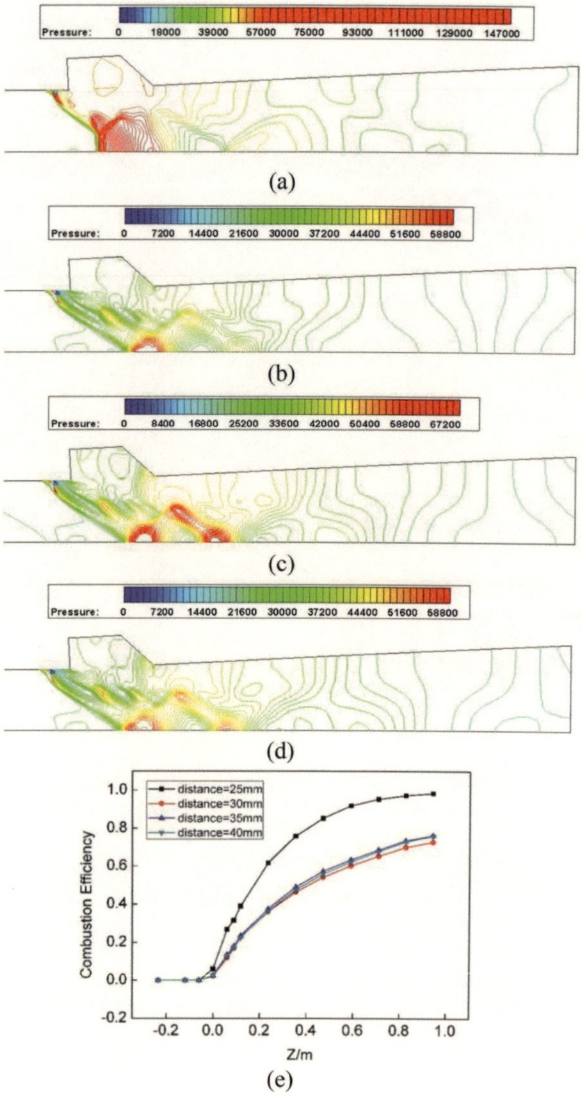

Figure 15.

Pressure contour (unit of P_a) and combustion efficiency for four fuel injection positions. (a) Distance between injection nozzle and cavity leading edge of 25 mm. (b) Distance between injection nozzle and cavity leading edge of 30 mm. (c) Distance between injection nozzle and cavity leading edge of 35 mm. (d) Distance between injection nozzle and cavity leading edge of 40 mm. (e) Comparison on the combustion efficiency.

The cavity has a great influence on the performance of supersonic combustors, such as combustion efficiency, drag characteristics, and flame stability. The impact of the cavity parameter variation on the performance of combustors is complex coupled. A surrogate model-based optimization and parameter analysis of the cavities in three-dimensional supersonic combustors with transverse fuel injection upstream were performed. The length, depth, and sweepback angle of cavities were first designed by orthogonal experiment. Numerical simulations were applied to analyze the performance and flow fields of the test cases. Surrogate models of the combustion efficiency and total pressure recovery coefficient with the design variables were constructed.

Based on the complex system optimization strategy, optimization of the cavity parameters was carried out twice to provide the Pareto front by the non-dominated

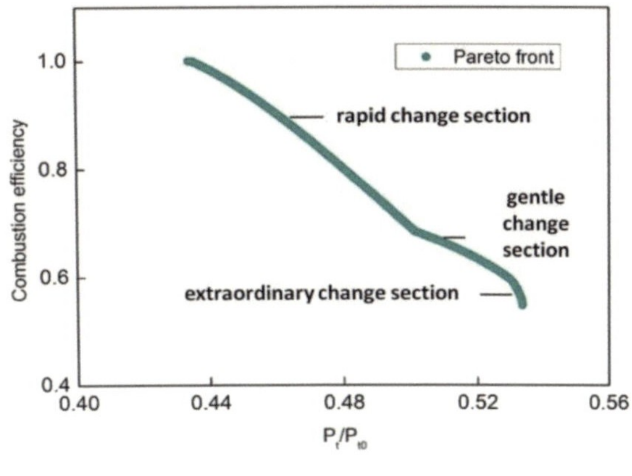

Figure 16.
Pareto front of the cavity optimization.

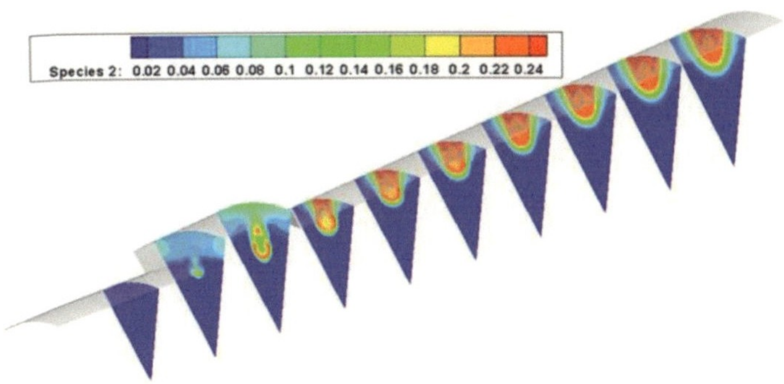

Figure 17.
Contour of the mass fraction of H_2O at different cross sections.

sorting genetic algorithm (NSGA-II). The results show that the optimal cavity configurations can be divided into narrow deep type, as can be seen in **Figure 16**, shallow long type, and medium deep and long type, which correspond to rapid change section, gentle change section, and extraordinary change section in the Pareto front. The combustion efficiency has a negative correlation with the length of cavities and a positive correlation with the depth of cavities, whereas the total pressure recovery coefficient has the opposite situations. Both combustion efficiency and total pressure recovery coefficient have few positive correlations with the sweepback angle. The combustors in the gentle change section have more uniform pressure distribution and higher total pressure recovery coefficient, which should be preferred when there is no need of high combustion efficiency. Optimized combustor configurations were simulated and verified compared to the baseline design, as shown in **Figure 17**.

3.4 Nozzles

A supersonic nozzle design is a significant work for hypersonic vehicles, which devotes to produce most of the thrust force and helps to improve the vehicle's internal/external integral level. Two-dimensional (2D) and axisymmetric minimum length nozzles (MLNs) with constant and variable specific heat were designed using the method of characteristics (MOCs) [20, 21], as can be seen in

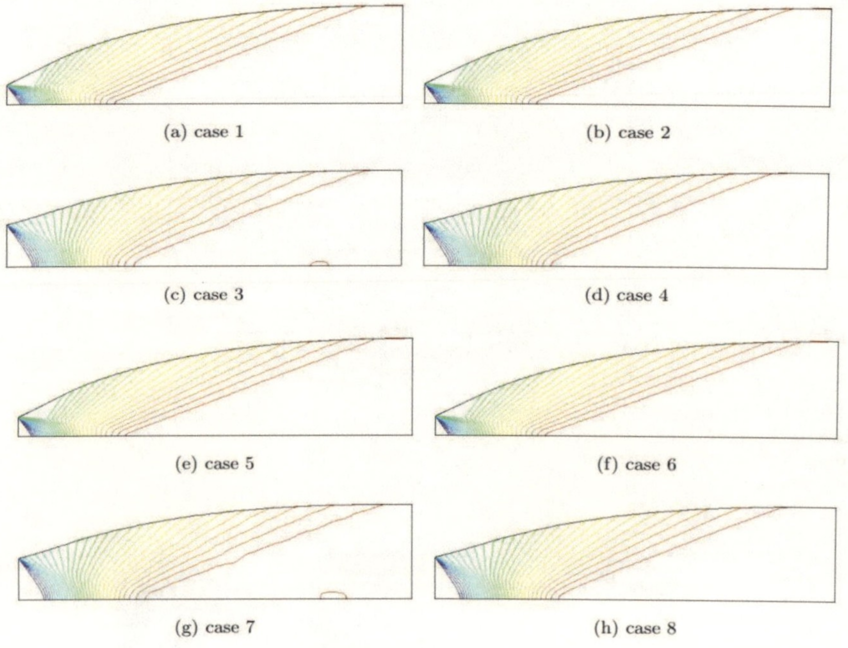

Figure 18.
Isograms of flow field Mach number in different nozzles.

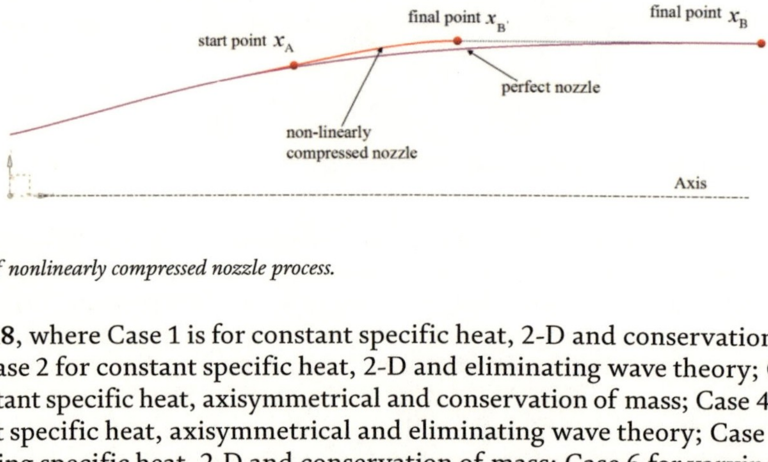

Figure 19.
Diagram of nonlinearly compressed nozzle process.

Figure 18, where Case 1 is for constant specific heat, 2-D and conservation of mass; Case 2 for constant specific heat, 2-D and eliminating wave theory; Case 3 for constant specific heat, axisymmetrical and conservation of mass; Case 4 for constant specific heat, axisymmetrical and eliminating wave theory; Case 5 for varying specific heat, 2-D and conservation of mass; Case 6 for varying specific heat, 2-D and eliminating wave theory; Case 7 for varying specific heat, axisymmetrical and conservation of mass; and finally, Case 8 for varying specific heat, axisymmetrical and eliminating wave theory. MOC is a numerical technique which has great advantages in accuracy and efficiency for solving hyperbolic partial differential equations.

After the two-dimensional supersonic nozzle design is finished, the flow field is simultaneously obtained with the MOC solution. Concerning the three-dimensional nozzle design, using the streamline tracing technique, the present work designed a three-dimensional asymmetric nozzle with a pre-determined offset circular entrance. However, the nozzles designed by MOC may have excellent thrust performance, but the length goes beyond the geometry constraints of the scramjet engine and does not meet the trim and lift-to-drag ratio requirements. As shown in **Figure 19**, a nonlinear compression technique can be used to truncate the streamline of the perfect nozzle, by preserving the initial major expansion parts of nozzle

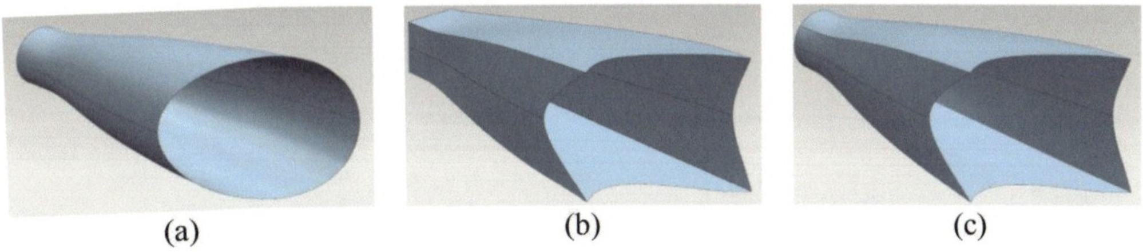

Figure 20.
Three-dimensional nonlinear truncated streamline traced nozzle: (a) circular, (b) rectangle, and (c) circular to rectangle.

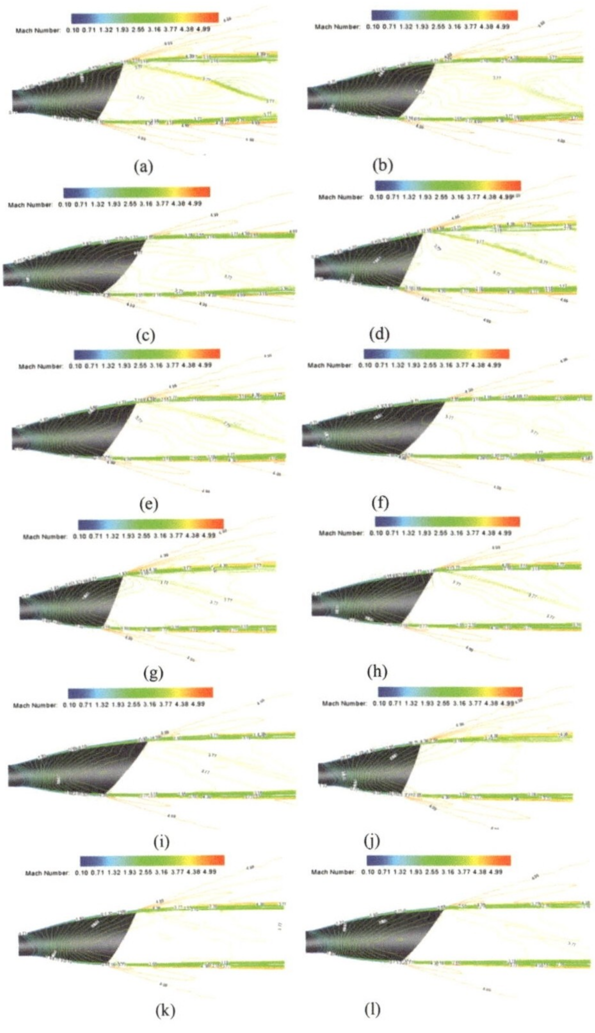

Figure 21.
Contour of Mach number under different truncation parameters. (a) 0.10–0.50 (b) 0.10–0.60 (c) 0.10–0.70 (d) 0.20–0.50 (e) 0.20–0.60 (f) 0.20–0.70 (g) 0.30–0.50 (h) 0.30–0.60 (i) 0.30–0.70 (j) 0.40–0.50 (k) 0.40–0.60 (l) 0.40–0.70 (First parameter is $A = x_{B'}/x_{B}$ *and second one is* $B = x_{A}/x_{B}$*).*

and making the remaining nonlinear compressed. **Figure 20** shows three kinds of typical generated three-dimensional asymmetric nozzle. Numerical studies were performed to investigate the design parameters, such as the pre-determined offset size, the nonlinear truncation parameters, and so on. **Figure 21** shows the simulated Mach number contours with different truncation parameters.

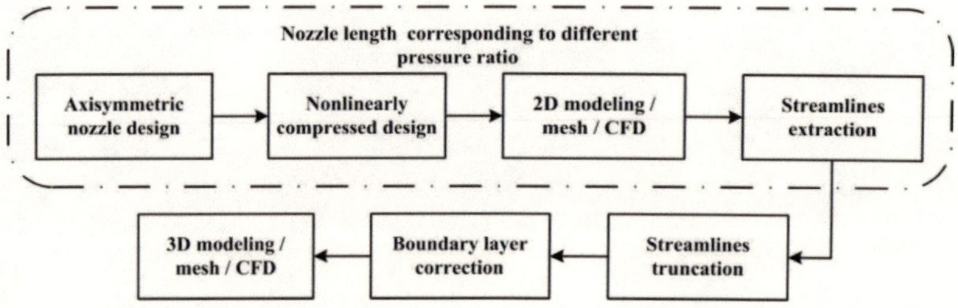

Figure 22.
Three-dimensional asymmetric nozzle design process.

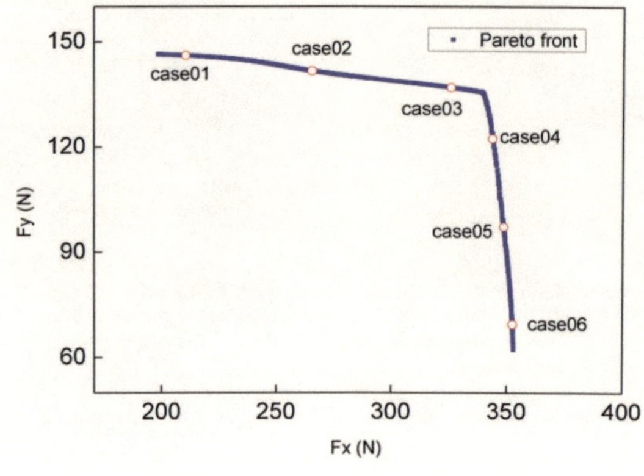

Figure 23.
Pareto front with six selected cases.

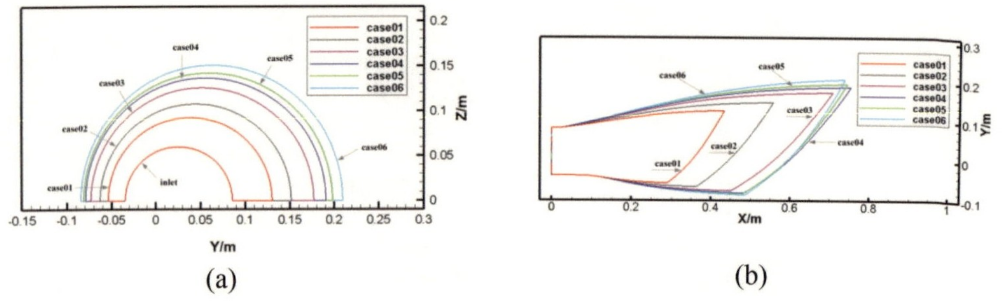

Figure 24.
Inlet and outlet shapes of six selected optimization: (a) inlet shape and (b) outlet shape.

Figure 22 shows the whole design process. Thereafter, a multiobjective design optimization has been performed, using orthogonal design, Kriging surrogate model with objective functions of thrust and lift force. Numerical simulations were conducted to validate the accuracy of surrogate models and to provide details of flow fields. The optimization results were examined to investigate the key factors and underlying flow physics that influenced the nozzle performance and to offer a preliminary guide to design a better nozzle with a suitable length. Six cases were selected from the Pareto front as illustrated in **Figures 23** and **24**. The design variables of these cases are used to obtain new design results. New numerical simulations were performed to verify the reliability of surrogate models and to provide a deeper insight of nozzle performance, as can be seen in **Figure 25**.

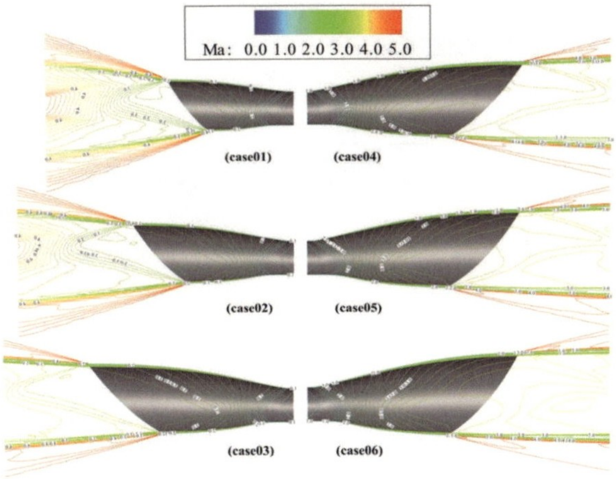

Figure 25.
Mach number contour of six Pareto front cases.

4. Airframe-propulsion integration

The integration method for the cone-derived waverider and scramjet is introduced as the following. The basic Busemann flow field is decided according to the area ratio between inlet and outlet surfaces of the inward-turning inlet, which can be calculated using the stream thrust analysis. As a result, the Busemann inlet can be obtained. The length of the Busemann inlet is too large, and its upper surface of inlet is horizontal which is not appropriate for integration with the waverider. The Busemann inlet is then truncated.

The angle of the truncation cone should not be too large and is chosen between 3 degree and 5 degree in the present studies. Meanwhile, the semiapex angle of the truncation cone should be smaller than that of the waverider. Accordingly, iterations are necessary to find out the appropriate basic Busemann flow field. The basic flow field of the cone-derived waverider is decided with the design parameters, for example, the inflow Mach number and an appropriate compression angle. The shock angle and the semiapex angle can be calculated. As mentioned before, the semiapex angle must be larger than that of the truncated Busemann inlet. **Figure 26** shows the integration of the inlet and the waverider compression surface. An integration example of the cone-derived waverider and the scramjet can be seen in **Figure 27**.

Figure 28 compares the contour of Mach number of original cone-derived waverider and integrated vehicle. It can be observed that shock waves attach the bottom leading edges of surfaces of both vehicles. **Figure 29** shows simulation

Figure 26.
Integration of cone-derived waverider and truncated Busemann inlet.

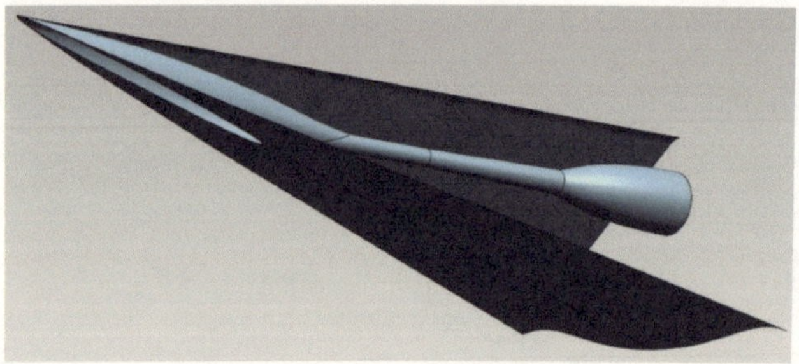

Figure 27.
Integration vehicle.

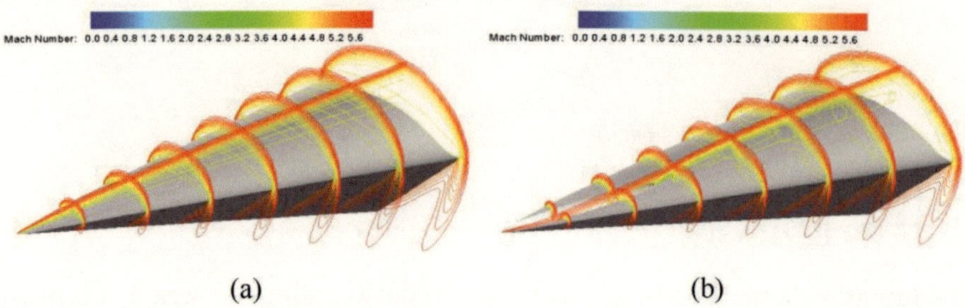

Figure 28.
Comparison on simulated Mach contour of original waverider and integration vehicle. (a) Original waverider. (b) Integration vehicle.

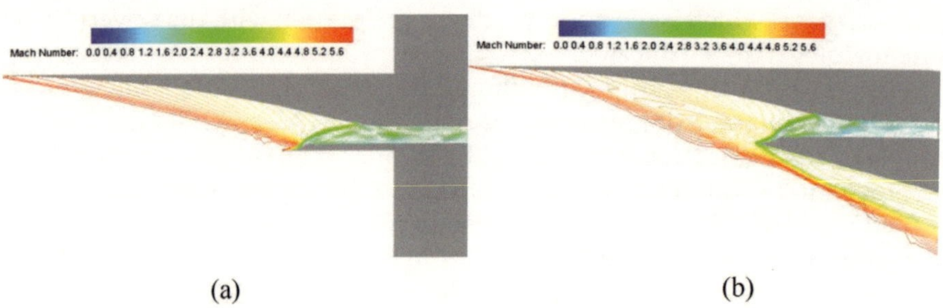

Figure 29.
Comparison on simulated Mach contour of original Busemann inlet and integration vehicle. (a) Original Busemann inlet. (b) Integration vehicle.

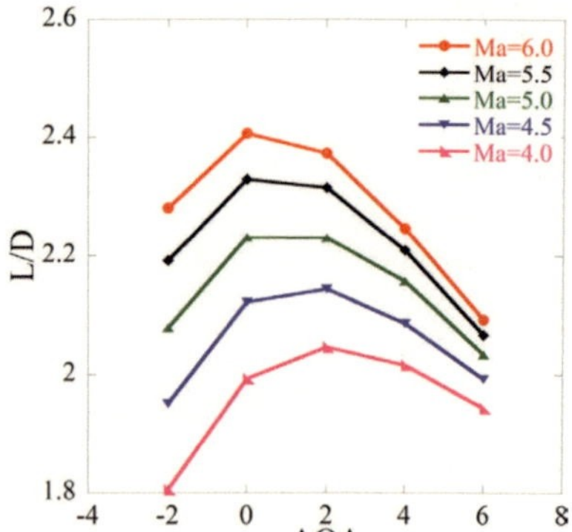

Figure 30.
Lift-to-drag ratio versus angle of attack of the integration vehicle.

results of the original truncated Busemann inlet and the integrated vehicle. There is a small difference between them since the integration would change the inflow for the inlet. However, the difference is not large. **Figure 30** shows the lift-to-drag ratio versus angle of attack of the integration vehicle.

5. Conclusions

Airframe-propulsion integration design method is investigated in the present study. The design methods for the waverider and each components of the scramjet are introduced. The integration method between the waverider and the scramjet is described. The overall optimization for the whole scramjet flowpath is optimized with quick engineering estimation method to provide appropriate performance criteria, which are then used to design three-dimensional component configurations.

Additionally, optimization is performed for the waverider and scramjet components with surrogate modes and CFD simulations. Numerical studies are carried out to find out the performances of the waverider and each component of the scramjet to check whether they can work normally under the design conditions. In the future, the design method for the dual-mode combustor (ramjet and scramjet) will be considered. The numerical simulation for the whole scramjet or dual-mode combustor is necessary to perform to verify the design method.

Conflict of interest

The authors declare that there is no conflict of interest regarding the publication of this work.

Author details

Yao Zheng*, Shuai Zhang, Tianlai Gu, Meijun Zhu, Lei Fu, Minghui Chen and Shuai Zhou
Center for Engineering and Scientific Computation, and School of Aeronautics and Astronautics, Zhejiang University, Hangzhou, Zhejiang, China

*Address all correspondence to: yao.zheng@zju.edu.cn

References

[1] Huang W, Wang ZG. Numerical study of attack angle characteristics for integrated hypersonic vehicle. Applied Mathematics and Mechanics. 2009;**30**(6):779-786. DOI: 10.1007/s10483-009-0612-y

[2] O'Neill MK, Lewis MJ. Optimized scramjet integration on a waverider. Journal of Aircraft. 1992;**29**(6):1114-1121. DOI: 10.2514/3.56866

[3] Huang W, Wang ZG, Jin L, Liu J. Effect of cavity location on combustion flow field of integrated hypersonic vehicle in near space. Journal of Vision. 2011;**14**(4): 339-351. DOI: 10.1007/s12650-011-0100-3

[4] Stevens DR. Practical Considerations in Waverider Applications. AIAA Paper 92-42471992. DOI: 10.2514/6.1992-4247

[5] Takashima N, Lewis MJ. Waverider Configurations Based on Nonaxisymmetric Flow Fields for Engine-Airframe Integration. AIAA Paper 94-03801994. DOI: 10.2514/6.1994-380

[6] Takashima N, Lewis MJ. Engine-Airframe Integration on Osculating Cone Waverider-Based Vehicle Designs. AIAA Paper 96-25511996. DOI: 10.2514/6.1996-2551

[7] O'Brien TF, Lewis MJ. Rocket-based combined-cycle engine integration on an osculating cone waverider vehicle. Journal of Aircraft. 2001;**38**(6):1117-1123. DOI: 10.2514/2.2880

[8] Ji T, Wang Y, Zhang J, Zhang S. Surrogate models characteristic analysis and optimization for waverider aerodynamic design (in Chinese). Journal of Solid Rocket Technology. 2015;**38**(4):451-457. DOI: 107673/j.issn.1006-2793.2015.04.001

[9] William H, Pratt H, David T. Hypersonic Airbreathing Propulsion. 5th ed. Washington, DC: American Institute of Aeronautics and Astronautics Inc; 1994. 173 p

[10] Birzer CH, Doolan CJ. Quasi-one-dimensional model of hydrogen-fueled scramjet combustors. Journal of Propulsion and Power. 2009;**25**(6):1220-1225. DOI: 10.2514/1.43716

[11] Zhu M, Gu T, Zhang S, Zheng Y. Optimization and parameter analysis of cavity in a three-dimensional supersonic combustor (in Chinese). Journal of Propulsion Technology. 2018;**39**(8):1780-1789. DOI: 10.13675/j.cnki.tjjs.2018.08.012

[12] Gu T, Zhang S, Zheng Y. Performance of a jaws inlet under off-design conditions. Proceedings of the Institution of Mechanical Engineers, Part G: Journal of Aerospace Engineering. 2017;**231**(2):294-305. DOI: 10.1177/0954410016636163

[13] Zhu M, Fu L, Zhang S, Zheng Y. Design and optimization of three-dimensional supersonic asymmetric truncated nozzle. Proceedings of the Institution of Mechanical Engineers, Part G: Journal of Aerospace Engineering. 2018;**232**(15):2923-2935. DOI: 10.1177/0954410017718567

[14] Zhu M, Zhang S, Zheng Y. Conceptual design and optimization of scramjet engines using the exergy method. Journal of the Brazilian Society of Mechanical Sciences and Engineering. 2018;**40**(12):553. DOI: 10.1007/s40430-018-1468-y

[15] Jones JG, Moore KC, Pike J, Roe PL. A method for designing lifting configurations for high supersonic speeds, using axisymmetric flow fields. Ingenieur Archiv. 1968;**37**:56-72. DOI: 10.1007/BF00532683

[16] Kothari AP, Tarpley C, McLaughlin TA, et al. Hypersonic vehicle design using

inward turning flowfields. In: 32nd Joint Propulsion Conference and Exhibit, 1996-2552. 1996

[17] Balakrishnan G, Williams F. Turbulent combustion regimes for hypersonic propulsion employing hydrogen-air diffusion flames. Journal of Propulsion and Power. 1994;**10**(3):434-437. DOI: 10.2514/3.23754

[18] Smith S, Mungal M. Mixing, structure and scaling of the jet in crossflow. Journal of Fluid Mechanics. 1998;**357**:83-122. DOI: 10.1017/S0022112097007891

[19] Hasselbrink EF, Mungal M. Transverse jets and jet flames. Part 1. Scaling laws for strong transverse jets. Journal of Fluid Mechanics. 2001;**443**: 1-25. DOI: 10.1017/S0022112001005146

[20] Fu L, Zhang S, Zheng Y. Performances analysis of asymmetric minimum length nozzles. International Journal of Modeling, Simulation, and Scientific Computing. 2016;7(2):1-12. DOI: 10.1142/S1793962316500215

[21] Lei F, Zhang S, Zheng Y. Design and verification of minimum length nozzles with specific/variable heat ratio based on method of characteristics. International Journal of Computational Methods. 2016;**13**(6):1-13. DOI: 10.1142/S0219876216500341

Chapter 4

Inversely Designed Scramjet Flow-Path

Mookesh Dhanasar, Frederick Ferguson and Julio Mendez

Abstract

It can be argued that at the heart of functional hypersonic vehicle is its engine. Key to a functionally efficient scramjet engine lies in the design of its flow-path. The flow-path is made up of the following sections: (1) the forebody inlet; (2) the isolator, (3) the combustor, and (4) the nozzle. This chapter focuses on the design of the forebody inlet and the isolator sections of a scramjet engine. In this framework, key to a functionally efficient scramjet engine lies in the design of its flow-path. This flow-path design must consider a complex flow-field physics and the interaction of physical surfaces with this complex flow-field. Many attempts to design efficient scramjet flow-paths have met with some measured degree of success. This research uses a 'inverse design' approach, which is similar to Darwin's theory of evolution, where an organism adopts to survive in its environment; the scramjet flow-path will be carved/extracted from the operational environment. The objective is to naturally and organically capture, process and direct the flow from the environment; thus preparing it for the combustion process. This approach uses the ideal 2-D oblique shock relations, coupled with Nonweiler's caret waverider theory and streamline marching techniques.

Keywords: scramjet flow-path, hypersonic propulsion, inverse design, stream tube, oblique shockwave, Billig's isolator relations, Nonweiler's caret waverider

1. Introduction

Driven by the desire to improve air travel and shorten flight time, aircraft engines have evolved from simple reciprocating internal combustion engines to advance axil flow jet engines. Jet engines fall into several categories. These include air-breathing, turbine powered, turbojet, turbofan, ramjet compression and scramjet compression engines. Ramjet and scramjet compression engines are unique in that they represent the latest development on the evolutionary path of jet engines. The ramjet, unlike conventional jet engines which uses turbine driven compressors to compress the incoming air, uses shockwaves to achieve this goal. The compressed air is burnt in the combustor under sub-sonic conditions. The scramjet is basically an air-breathing jet engine designed to fly at hypersonic speeds between Mach 4 and 12 or speeds in the range of 1207–2995 m/s (2700–6700 mph). A scramjet engine captures its airflow from the atmosphere and also compresses it across shockwaves before the air enters the combustor. Fuel is injected into the combustor where combustion occurs under supersonic conditions. The hot, high-pressure gas leaving the combustor is then accelerated to high velocities in the nozzle to produce thrust as it exits the engine.

Generally speaking, the concepts associated with scramjet engines appear at first glance to be very simple. This however is very misleading as attempts develop a working scramjet engine that has proven to be quite an engineering challenge. Several aspects of scramjet engine development are at various stages of development. These include supersonic fuel-air mixing, aero-thermodynamic heat dissipation from both skin friction and internal combustion, and other thermal management problems associated with operating an engine at exceedingly high temperatures for extended periods of time. Combustion chamber components could experience temperatures on the order of over 3033 K (5000°F). At these temperatures most metals melt and fluids (air and fuel) ionize, making the physics of their associated behavior unpredictable.

This chapter focuses on the design concepts for the forebody, inlet, and isolator sections of an innovative scramjet engine geometry and some of its flow physics.

2. Inverse scramjet 2-D centerline design approach

As stated earlier, the scramjet concept represents the latest evolution in the series of air-breathing jet engines. Combustion in these engines occurs under supersonic conditions. Scramjet engines are seen as the propulsion system that is at the heart of hypersonic vehicles/platforms. Every scramjet conceptual engine design and engines flown to-date all have a common set of components or sub-sections. **Figure 1** presents these components/sub-sections for a pod-mounted conceptual scramjet design. These components/sections are the forebody section, the inlet section, isolator section, combustor section, and the diffuser-nozzle section. Ideally, the engine concept presented should be able to function over a wide range of Mach numbers. This gives rise to the idea of a morphing ramjet/scramjet or dual mode scramjet configurations as presented in **Figure 2** [1]. **Figure 2a**, presents the dual mode scramjet engine, **Figure 2b**, the pure scramjet mode and **Figure 2c**, the pure ramjet mode.

A typical dual mode scramjet configuration as that presented in **Figure 3**, was inversely carved out of supersonic and hypersonic flow-fields. The design framework used in the design of the forebody, inlet and isolator sections forms the core of this chapter.

3. Scramjet inverse design approach

The inverse design approach relies on extracting the configuration of interest from the environment in which it operates. For this design process the centerline

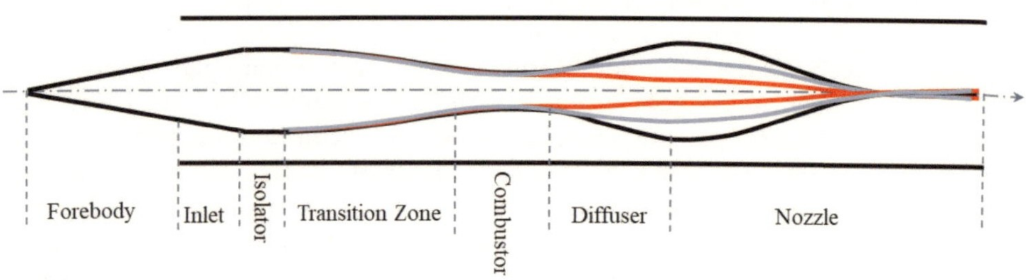

Figure 1.
Pod-mounted scramjet concept.

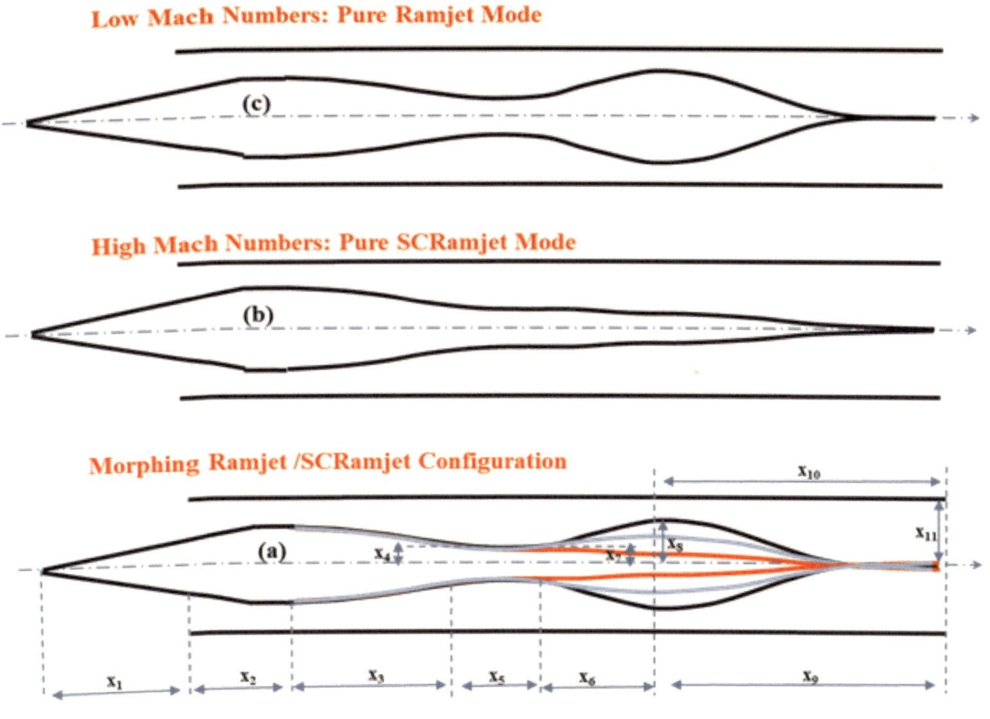

Figure 2.
Dual-mode scramjet concept.

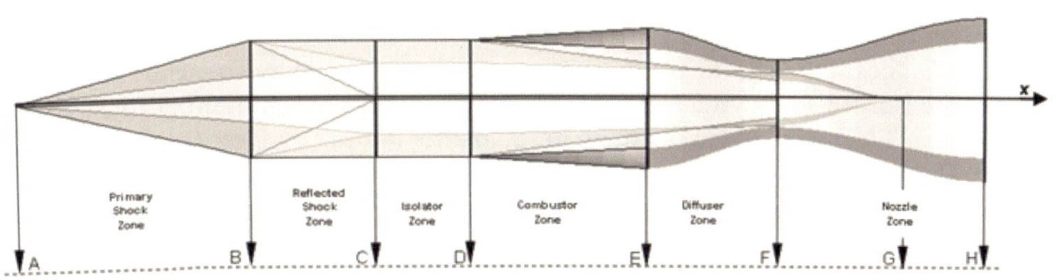

Figure 3.
Illustration of the cross section of the scramjet.

geometry of a given 2-D scramjet configuration is explicitly constructed using the following design inputs: freestream Mach number, M_∞, scramjet forebody length, L, shock angle, β, caret angle, α, cruising flight altitude, H_∞, and isolator back-pressure ratio, P_{in}/P_{exit}. All freestream flow-field properties are extracted from the Mach number and altitude [3–6]. This information is used in the construction, analysis and definition of the three fundamental aerodynamic zones, namely; the '*primary shock*' zone AB, the '*reflected shock zone*', BC, and the '*isolator zone*', CD as presented in **Figure 4**. Also presented in **Figure 4** is a 2-dimensional conceptual representation of the flow-field physics associated with supersonic flow interaction over a wedge and in a constant area duct. Details of this flow-field physics and its exploitation in the inverse design approach are explained in the next section. Also addressed is the derivation of the actual 3-dimensional forebody section. This is a

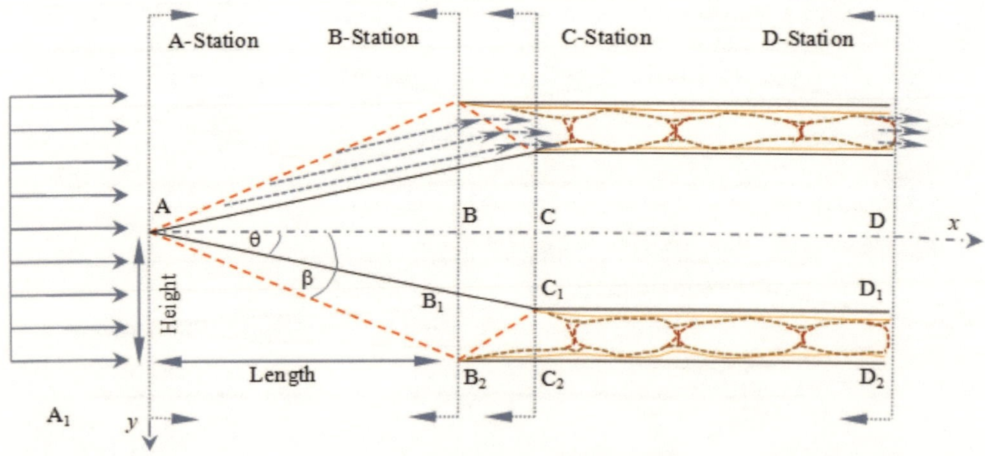

Figure 4.
Conceptual 2-D centerline cross-section of the forebody-inlet-isolator scramjet section with flow physics representation.

two-step process, where in step one, the 2-D construction of the 'forebody', domain *A-D*, is conducted. Step two is where the 3-D geometry is obtained.

3.1 Aerodynamics of the 2-D 'forebody' configuration

Consider the 2-D cross-sectional illustration of the scramjet forebody-inlet-isolator section presented in **Figure 4**. Now consider a supersonic flow travelling parallel to the x-axis of a 2-D wedge. Supersonic aerodynamics dictates that the flow is deflected first by the oblique shock wave, AB_2, originating from the leading edge, A, of the wedge. The flow is deflected a second time by a reflected shock wave, B_2C_1 emanating from the cowl lip at point B_2, of the inlet. The flow enters the isolator duct and travels once more in a direction that is parallel to the x-axis. To ensure that the flow in the isolator duct remain supersonic the freestream Mach number must be greater than 3.0 and the shock wave angle, β, greater than 12 and less than 30 degrees.

The flow-field behavior within the isolator duct is of paramount importance. This flow-field may consist of a system of oblique or normal shocks, as visualized in **Figure 4**. Driving this behavior is the flow-field vicious interactions with the isolator duct walls. The isolator's non-dimensional length, L/H, and the pressure differential at the duct's entrance and exit also enhance the flow-field's behavior.

3.2 Derivation of the 2-D 'forebody-inlet-isolator' configuration

The 'forebody-inlet-isolator' concept presented in **Figure 4** relies on determining the geometric design points located at stations A, B, C and D, along the x-axis of the scramjet. This is accomplished by use of the oblique shock relations described in [2–7] and the 'isolator' relations that were experimentally derived in [8–9]. It is assumed that in **Figure 4** the flow travels in the x-direction, and that the construction of the 'forebody' configuration starts at design point, A. The following account details the logic used to define the locations of design points A, B, C and D:

3.2.1 Design point at station A

The design point at station A is considered the origin of the scramjet design coordinate system, therefore, design point A coordinates are evaluated as follows, $A_x = 0$, $A_y = 0$, and $A_z = 0$.

3.2.2 Design points at station B

Using the input data, the location of design point B, can be computed with the use of the following relations: $B_x = L$, $B_y = 0$, and $B_z = 0$.

In addition, using trigonometric relationships design point B_1 is evaluated as follows: $B_1x = B_x$, $B_{1y} = B_x\tan(\theta)$ and $B_{1z} = 0$.

The coordinates for design point B_2 are evaluated in the following manner: $B_{2x} = B_x$, $B_{2y} = B_x\tan(\beta)$, and $B_{2z} = 0$. The wedge angle is represented by theta (θ) and the shock angle is represented by beta (β). Using the Mach number and the shock angle beta (β), the wedge angle theta (θ) can be obtained with the use of the Theta-Beta-Mach (θ-β-M) relationship [2–7] given as seen in Eq. (1). In Eq. (1) the constant γ is set at a value of 1.4.

$$\theta = \operatorname{atan}\left\{2\cot\beta\left[\frac{M_\infty^2 \sin^2\beta - 1}{M_\infty^2(\gamma + \cos 2\beta) + 2}\right]\right\} \tag{1}$$

3.2.3 Design points at station C

The design points at station C is extracted from the wedge angle, θ, and the flow-field properties behind the primary shock wave, AB_2, as seen in **Figure 4**. Determination of the location of design point C is a little more involved and is approached systematically as outlined in the following steps:

a. First, the Mach number, M, behind the primary shock wave, AB_2 (see **Figure 4**), is obtained using Eq. (2),

$$M = \left\{\frac{1}{\sin(\beta - \theta)}\right\}\frac{\left[1 + [(\gamma - 1)/2](M_\infty \sin(\beta))^2\right]}{\left[\gamma(M_\infty \sin(\beta))^2 - (\gamma - 1)/2\right]} \tag{2}$$

b. This Mach number, coupled with the free stream parameters are then used with the oblique shock relations derived in [5] for the evaluation of all of flow-field properties behind the primary shock, AB_2. The flow-field properties, pressure, P, temperature, T, density, ρ, and total pressure, $P_{t,2}$, are evaluated using Eqs. (3)–(6).

$$\frac{P}{P_\infty} = \frac{2\gamma(M_\infty \sin\beta)^2 - (\gamma - 1)}{(\gamma + 1)} \tag{3}$$

$$\frac{T}{T_\infty} = \frac{\left[2\gamma(M_\infty \sin\beta)^2 - (\gamma - 1)\right]\left[(\gamma + 1)(M_\infty \sin\beta)^2 + 2\right]}{(\gamma + 1)^2(M_\infty \sin\beta)^2} \tag{4}$$

$$\frac{\rho}{\rho_\infty} = \frac{(\gamma + 1)(M_\infty \sin\beta)^2}{(\gamma - 1)(M_\infty \sin\beta)^2 + 2} \tag{5}$$

$$\frac{P_{t,2}}{P_{t,\infty}} = \left[\frac{(\gamma + 1)(M_\infty \sin\beta)^2}{(\gamma - 1)(M_\infty \sin\beta)^2 + 2}\right]^{\frac{\gamma}{\gamma-1}}\left[\frac{(\gamma + 1)}{2\gamma(M_\infty \sin\beta)^2 - (\gamma - 1)}\right]^{\frac{1}{\gamma-1}} \tag{6}$$

c. B_2C_1 as seen in **Figure 4** represent the reflected shock wave. This reflected shock wave is a the result of a flow-field behind the primary shock wave, AB_2, with a supersonic Mach number, M, once more being deflected by an imaginary wedge, with wedge angle θ at design point B_2. This imaginary wedge is oriented in such a manner that it ensures that the deflected flow travels parallel to the x-axis, **Figure 4**. At this stage updated values for the wedge angle, θ and the Mach number, M, are obtained using Eqs. (1) and (2). A reflection shock angle, ϕ is now be defined as $\phi = \beta_1 - \theta$. In this expression, β_1 is the reflected shock angle. This reflected shock angle is generated by the interaction of the flow-field with Mach number M and the imaginary wedge with angle θ Note that β_1 is obtained using Eq. (1) and replacing the value of the freestream Mach number, M_∞, with that of the supersonic Mach number, M.

d. The flow-field properties behind the reflected shock wave B_2C_1 are now obtained in a similar manner as described in 'b' above. Eq. (2) is used to obtain M_1, which is the Mach number behind the reflected shock. In Eq. (2) the freestream Mach number, M_∞, is replaced with Mach number M. It is very important to note here that M_1 represents the Mach number at the entrance to the isolator section of the scramjet. Eqs. (3)–(6) are used to derive the additional flow-field properties of pressure, temperature, density and total temperature, p_1, T_1, ρ_1 and T_o, behind the reflected shock. Note that in these equations the value for the freestream Mach number, M_∞, is now replaced with the value of the Mach number, M, from the flow-field properties behind the primary shock.

e. Having obtained the parameters, θ, β and β_1 all design points at station C can now be derived. The y-coordinate and z-coordinate are defined as $C_y = 0$, and $C_z = 0$, respectively. The x-coordinate is obtained with the help of trigonometric relations, and is defined as:

$$C_x = \left[1 + \frac{\tan(\beta) - \tan(\theta)}{\tan(\theta) - \tan(\beta_1 - \theta)}\right] B_x \tag{7}$$

f. The coordinates of point C_1 are determined as follows: $C_{1x} = C_x$, $C_{1y} = C_x \tan(\theta)$, and $C_{1z} = 0$.

g. Similarly, the coordinates of point C_2 are determined from: $C_{2x} = C_x$, $C_{2y} = B_{2y}$, and $C_{2z} = 0$.

3.2.4 Design points at station D

The evaluation of the coordinates of the design points at station D is also a multi-step process.

a. First a non-dimensional expression for the '*normal total*' pressure value, $P_{n,in}$, is derived, Eq. (8). This expression is a function of isolator entrance conditions, where M_1 is treated as the M_{in}, and the static pressure, P_1, as P_{in}. Note here that the values of M_1 and P_1 are obtained from the flow-field properties behind the reflected shock B_2C_1.

$$\frac{P_{n,in}}{P_{in}} = \left[\frac{2\gamma M_1^2 - (\gamma - 1)}{(\gamma + 1)}\right] \tag{8}$$

In determining the isolator length for a design process, the ratio of the entrance to exit pressures, P_{in}/P_{out}, over the range between P_{in} and $P_{n,in}$ has to be evaluated. This value is needed to determine the length of an isolator that can reliably prevent all 'unstart' conditions. In this design process, the ratio, $P_{out}/P_{n,in}$, representing the isolator exit pressure, P_{out}, to the 'normal total' pressure value, $P_{n,in}$, is prescribed. Using this approach, the value for P_{in}/P_{out} can be determined by using Eq. (9):

$$\frac{P_{out}}{P_{in}} = \left(\frac{P_{out}}{P_{n,in}}\right)\left(\frac{P_{n,in}}{P_{in}}\right) \tag{9}$$

b. The system of 1-D conservation laws result in the following expression for the isolator exit Mach number, M_{out} [8, 9];

$$M_{out} = \left\{\frac{\gamma^2 M_{in}^2\left[1+((\gamma-1)/2)M_{in}^2\right]}{\left(1-\gamma M_{in}^2 - P_{out}/P_{in}\right)^2} - \left(\frac{\gamma-1}{2}\right)\right\}^{-\frac{1}{2}} \tag{10}$$

Similarly, with the exit Mach number known, the non-dimensional length of the isolator can be evaluated based on the following experimental relationship developed in [8, 9]:

$$\left(\frac{L}{H}\right)_{Isolator} = \frac{\sqrt{\theta/H}}{(\mathrm{Re}_\theta)^{\frac{1}{4}}}\frac{\left\{50((P_{out}/P_{in})+1)+170((P_{out}/P_{in})-1)^2\right\}}{M_{in}^2-1} \tag{11}$$

where Re_θ is the inlet Reynolds number based on the momentum thickness. Also, the symbol, H, represents the isolator height that is determined from the y-coordinates of points C_2 and C_1, in a manner such that, $H = C_{2y} - C_{1y}$.

c. The coordinates of point D are computed as follows: $D_x = C_x + L_{Isolator}$, $D_y = 0$, and $D_z = 0$.

d. The coordinates of point D_1 are computed as follows: $D_{1x} = D_x$, $D_{1y} = C_{1y}$, and $D_{1z} = 0$.

e. The coordinates of point D_2 are computed as follows: $D_{2x} = D_x$, $D_{2y} = C_{2y}$, and $D_{2z} = 0$.

Finally, with the coordinates of all the design points at all stations, A, B, B_1, B_2, C, C_1, C_2, D, D_1, and D_2, fully defined, the sketch illustrated in **Figure 4** can be constructed.

4. 3-D computer aided design (CAD) design

4.1 Overview of the 3-D design process

The 3-D design process has as its origin in the inversely design two-dimensional geometry extracted from a 2-D hypersonic flow-field. This is then coupled with the Nonweiler's waverider approach [10] of inversely carving stream surfaces from inviscid flow-fields. A caret waverider, **Figure 5**, is chosen as an example because it represents a 3-D geometry that was obtained from a 2-D flow-field. This caret waverider geometry is constructed from a single planer shock wave, AB_3B_4, as seen in **Figure 5**. A unique feature of this construction process is that at any cross-section of the waverider geometry there is a wedge that is supported by an oblique shock wave, with these wedges being parallel to the flow.

In reality, the caret waverider is carved from an inverse design approach that relies on the inviscid streamline principle. This principle states that any inviscid streamline can be replaced by a solid wall. The principle also states that replacing the inviscid streamline with a solid wall has no effect on the external flow. Planar inviscid stream surfaces are formed from these inviscid streamlines. These inviscid stream surfaces are then brought together to construct 3-D inviscid waverider geometries and stream tubes. An examination of **Figure 5** demonstrates how the streamlines form planar stream surfaces, such as, upper inviscid surfaces, ABB_3 and ABB_4, or lower stream surfaces, such as, AB_1B_3 and AB_1B_4.

This approach is further explained the next sub-section and is demonstrated by the construction of a supersonic 3-D wedge followed by a 3-D supersonic caret-shaped geometry. This caret-shaped geometry will then be used to generate supersonic star-shaped geometries of interest.

4.1.1 The 2-D forebody construction (side view)

The review of begins with the construction of the supersonic wedge. Established ideal oblique 2-D shockwave relationships are used to construct the supersonic 2-D forebody. There are two ideal oblique shock relationships which can be used, the Theta-Beta Mach relationship, or the Beta-Theta Mach relationship. In this review, the Theta-Beta Mach [3–5] relationship, described in Section 3.2 above is used in the construction of the supersonic 2-D forebody. For a prescribed Mach number, shock angle, Beta, at a given altitude, a wedge angle, Theta, is extracted. The next step is to set a forebody length. Having all the geometric data the 2-D forebody with the attached shock is constructed as presented in **Figure 6**.

4.1.2 The 2-D inlet construction (side view)

The inlet construction is an extension of the 2-D forebody construction. The oblique shock AB hits the cowl lip at point B and is reflected as shown in **Figure 6**.

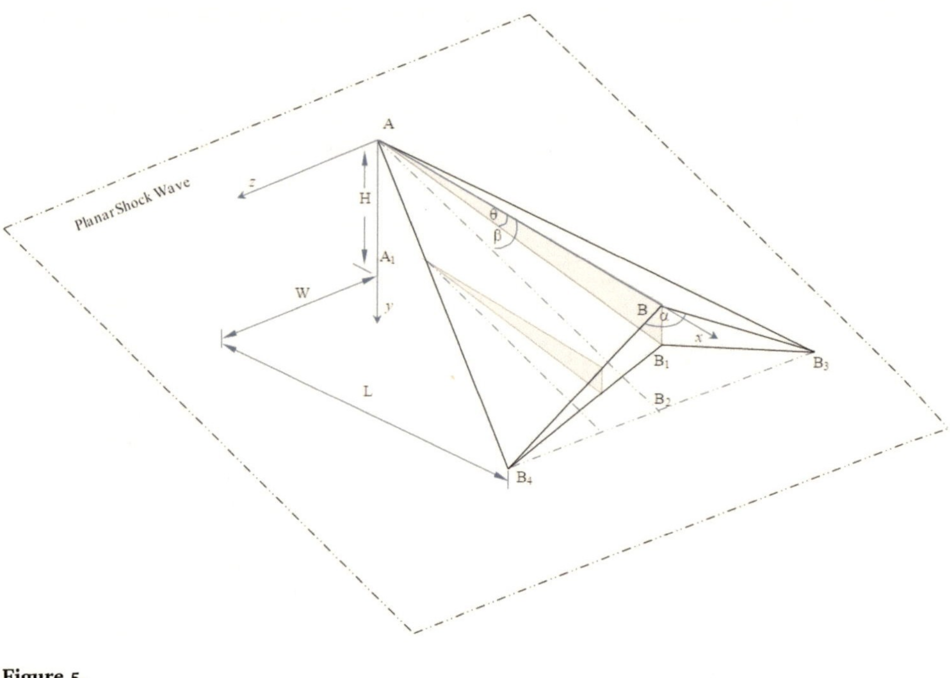

Figure 5.
Nonweiller caret wing waverider configuration.

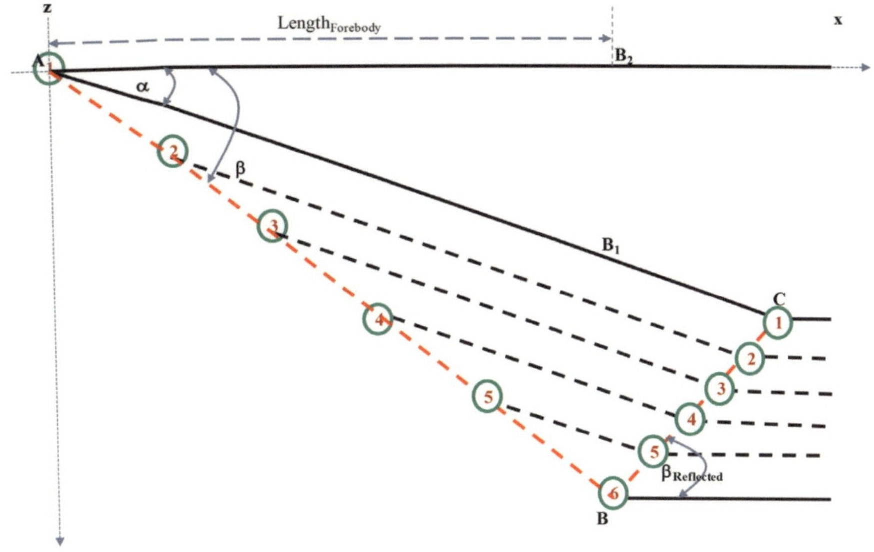

Figure 6.
Preparation for extracting information for 2-D base view.

Line BC represents the reflected shock from the interaction of the oblique shock and the cowl lip. Ideal oblique shock relationships are used to determine the reflected shock angle, Beta reflected. Note that the line AB_1 which represents the lower surface of the forebody continues to point C where it intersects with line BC. At this point in the design process the 2-D forebody and the inlet are constructed.

4.1.3 Streamline preparation of flow-field

A 2-D base view of the forebody-inlet components are constructed from geometric information obtained from the 2-D side view. The streamline cross-marching method used preserves both the geometric information and the 2-D flow-field information. The oblique shockwave, line AB, is first divided up into *N* number of equal parts, in this case six, as seen in **Figure 6**. Streamlines are then constructed emanating from the oblique shockwave. Each streamline has a starting point on the oblique shockwave, and ends on the reflected shockwave, line BC as presented in **Figure 6**. The longest streamline is represented by line AC and is the lower surface of the forebody-inlet. The shortest streamline is represented by point 6; here the streamline starts and stops at the same point. The streamlines emanating from the oblique shockwave and ending on the reflected shockwave travel parallel with respect to the lower surface of the forebody-inlet as presented in **Figure 6**. All streamlines are now processed by the reflected shockwave, BC, and travel parallel to the surfaces beginning at points C and B as shown in **Figure 6**. The 2-D base view can be extracted from the flow field.

4.1.4 Wedge geometry extraction from flow-field

The base view for the 2-D wedge is now extracted for the 2-D forebody-inlet and the associated 2-D flow-field. A zy-coordinate system is set up and a wedge width is prescribed. Streamlines emanating from the reflected shockwave, BC, are now mapped onto the zy-coordinate system as presented in **Figure 7**. Having completed the construction of the 2-D side view and the 2-D base view, the designer now has

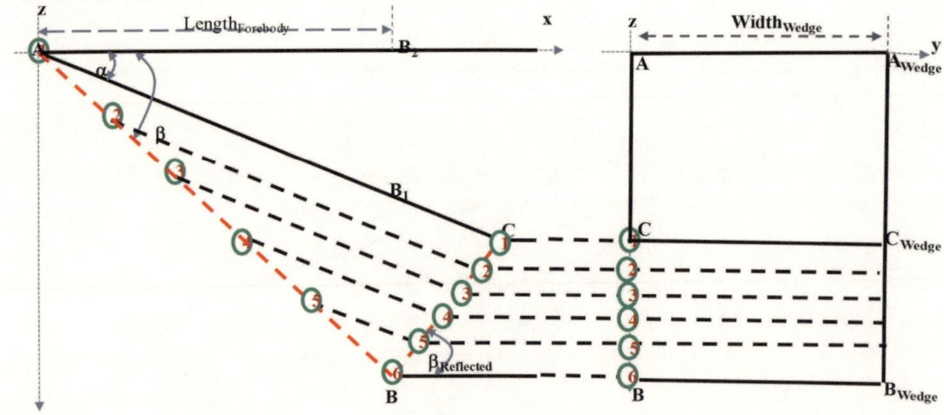

Figure 7.
Generation of 2-D base view for a wedge.

the 3-D coordinates that can be used to generate the 3-D forebody-inlet geometry for a 3-D wedge.

4.2 The caret geometry

The caret geometry forms the basis of the design of the star shaped geometries in this study. A similar process is used to obtain the caret-shaped 2-D base view. Now instead of providing a wedge width, a star angle, Phi, is provided as presented in **Figure 8**. For the four-point-star, Phi is 45 degrees. Reflecting points $AB_{Point}C$, about the z-axis will generate the 2-D base view for the caret-shaped waverider geometry. As before, all data required for the 3-D construction of the 3-D caret-shaped forebody-inlet have been extracted from the flow-field. **Figures 9–12** present the 3-D caret-shaped geometry obtained by using the design process described above and programmed using FORTRAN90/95.

4.3 3-D stream tube construction using the waverider approach

The scramjet forebody-inlet-isolator design concept as being proposed suggests a new use for waverider geometries. Here, the focus is not only of the waverider

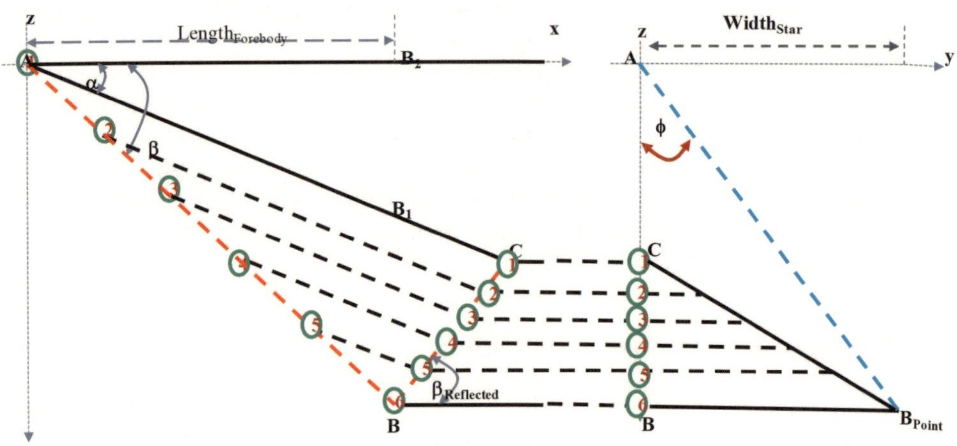

Figure 8.
Generation of 2-D caret-shaped geometry, base view.

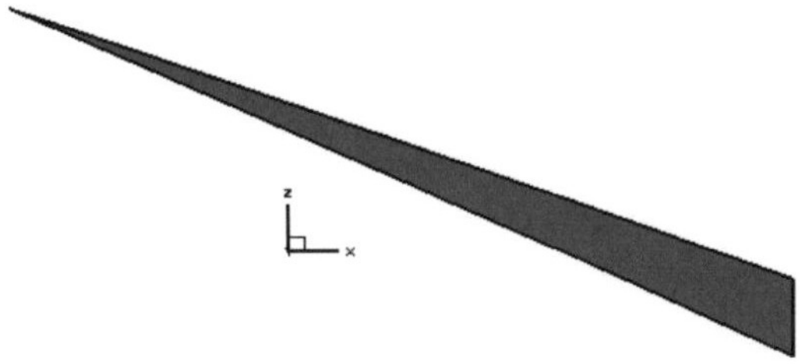

Figure 9.
Caret 2-D side view.

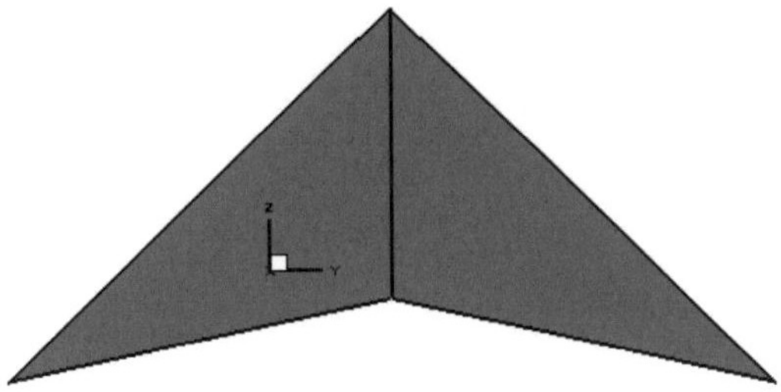

Figure 10.
Caret 2-D base view.

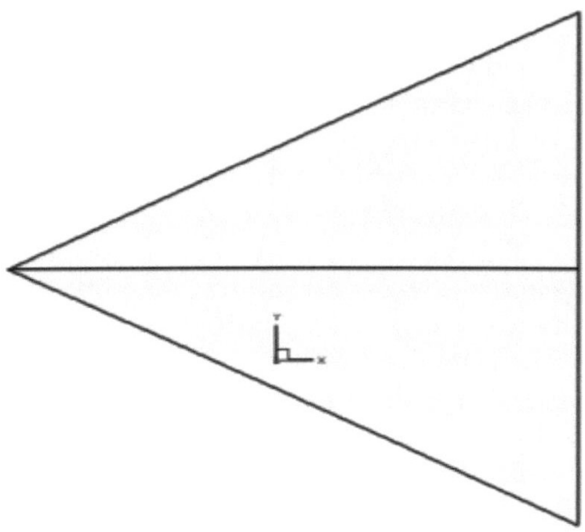

Figure 11.
Caret 2-D plan view.

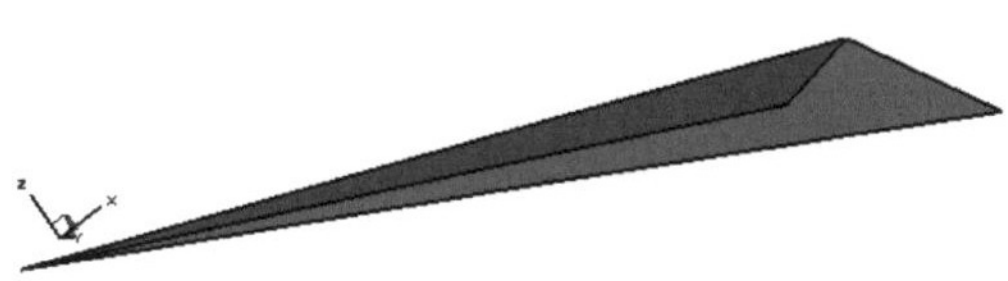

Figure 12.
Caret 2-D isometric view.

shape, but also on the external flow-field supporting the waverider configuration. As seen in **Figure 5**, attention is on the external 2-D flow on the waverider lower surfaces, that is, AB_1B_3 and AB_1B_4, and the flow entering and exiting the planes, AB_3B_4 and $B_1B_3B_4$. With this alternative perspective, the innovation lies in the fact that the flow moving across the lower surface of the waverider is treated as the flow

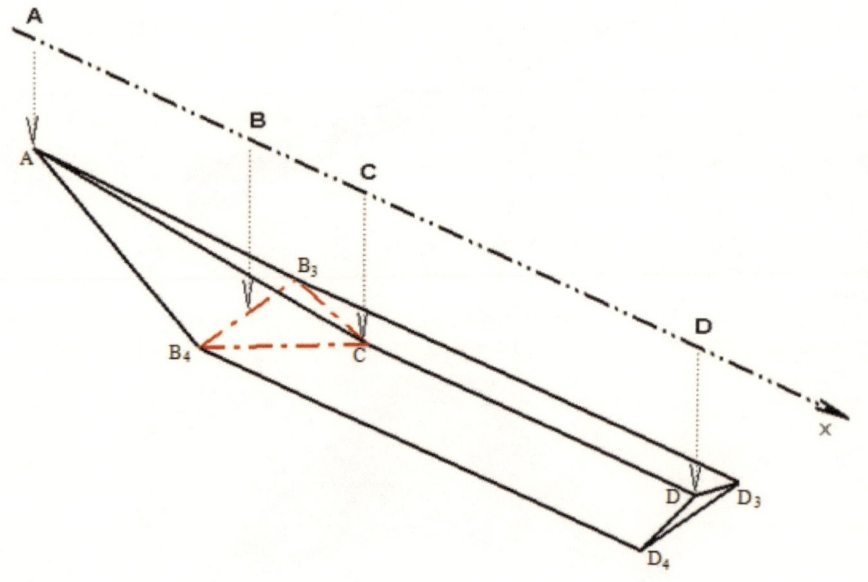

Figure 13.
Waverider derived stream tube.

entering a stream tube through surface, AB_3B_4 and leaving through the plane, $B_1B_3B_4$. Recall at this point that the flow-field is two dimensional, confined to the xy-plane and can be treated as a collection of 2-D slices that are parallel to each other. The flow within the stream tube is bounded by the lower inviscid surfaces, AB_1B_3 and AB_1B_4 and an imaginary line surface, B_3B_4.

A completed stream tube consisting of the forebody-inlet-isolator sections is presented in **Figure 13**. This stream tube is carved/extracted from a supersonic flow-field travelling parallel to the x-axis, which is compressed by two oblique shock waves; resulting in the flow once again traveling in a direction parallel to the x-axis. Further examination of **Figure 13** identifies the primary shock wave plane as AB_3B_4, which supports two compression surfaces, ACB_3 and ACB_4. At this stage the flow field is no longer parallel to the x-axis. A reflected shock wave is constructed to form the plane, CB_3B_4. This specially designed plane, CB_3B_4, now straightens the flow leaving the shock surface, CB_3B_4, so that it once again travels parallel to the x-axis. The reflected flow now forms the stream tube comprising of the following planar surfaces, CDD_3B_3, CDD_4B_4, and $B_3B_4D_4B_3$.

4.4 Transforming stream tubes to 'star' shaped geometries

The preceding section saw the design of a single stream tube. These single stream tubes can now be used to create star-shaped geometries of interest, an example of which is presented in **Figure 14**. Presented in **Figure 14** is a four point star geometry, so termed because it is a collection of four stream tubes that is assembled in a manner to create a 'closed form' geometry of interest.

The fundamental concept in moving from a 2-D geometry, **Figure 4**, to the 3-D geometries, **Figures 13–16**, lies mainly on identifying the coordinates along the z-axis. Determination of the location of points, B_3, B_4, D_3 and D_4, is of significant importance. These points are responsible for the development of a closed form geometry/closed tube with the ability of preserving the aerodynamics associated with the inviscid flow-field behavior. Additionally, the 'y' and 'z' coordinates of these points rely of the choice of angle α, an example of which is the angle D_3DD_4 as

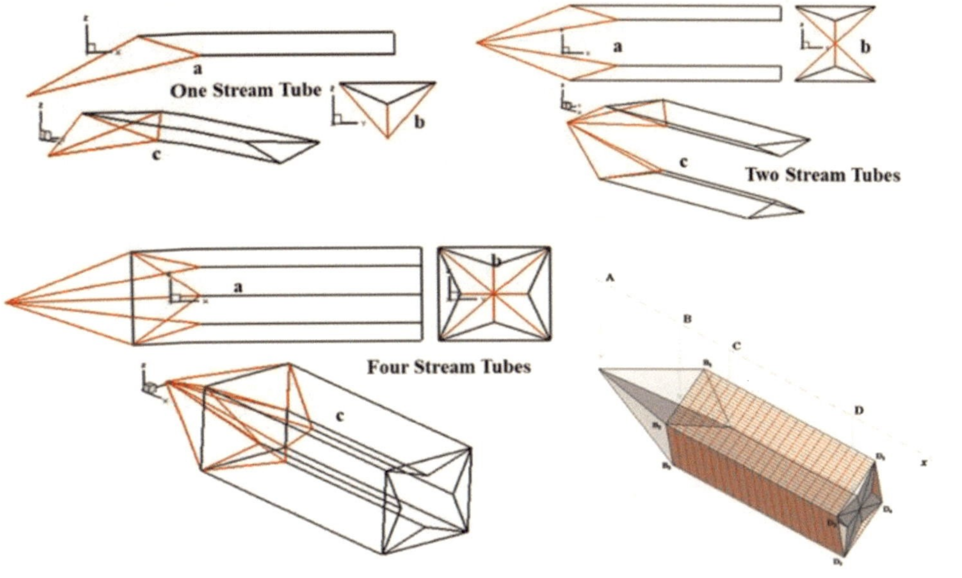

Figure 14.
A 4-points star-shaped scramjet forebody-inlet-isolator [1].

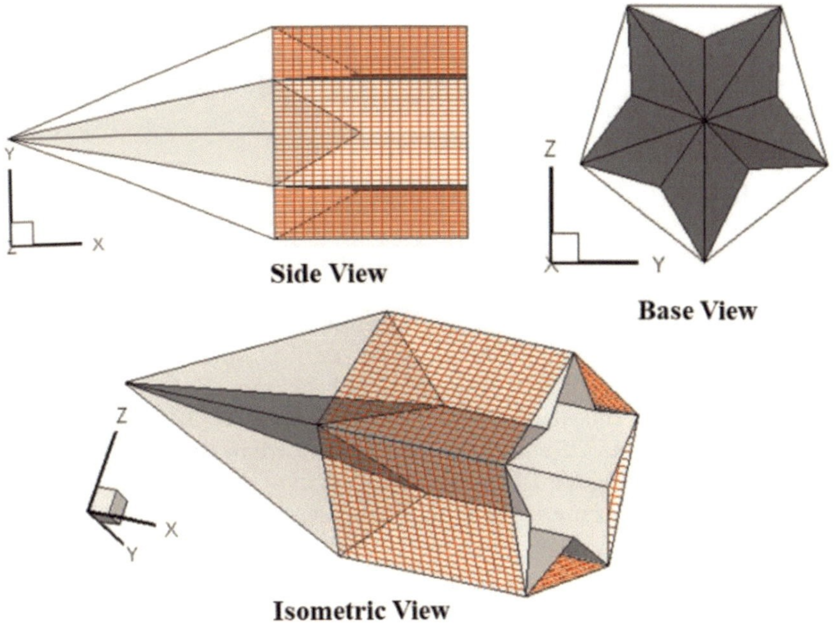

Figure 15.
Five-points scramjet forebody-inlet-isolator [1].

seen in **Figure 13**. In generating the four point star configuration the angle α is set to 90 degrees.

5. Validation section

This section focuses on the validation of the forebody-inlet-isolator sections associated with the proposed scramjet engine concept. The independent

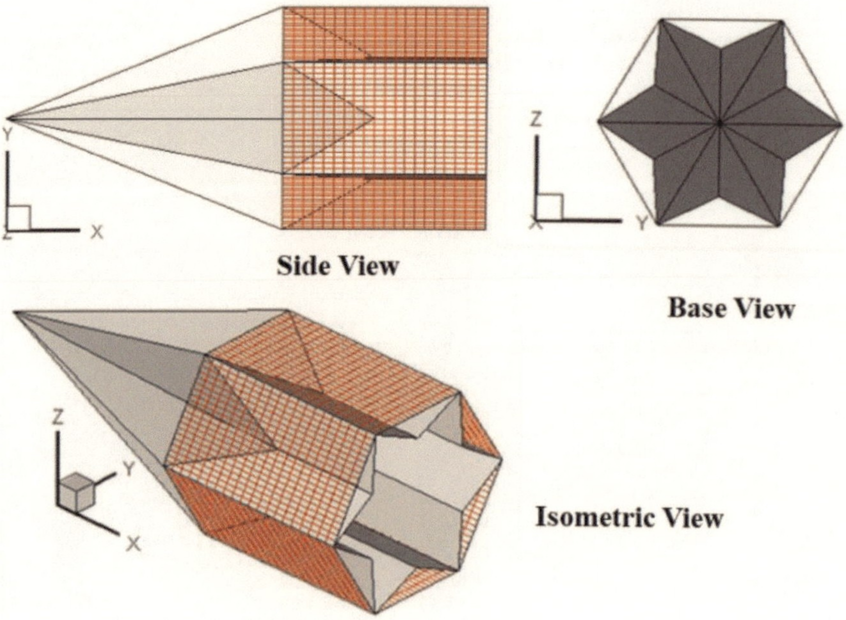

Figure 16.
Six-points scramjet forebody-inlet-isolator [1].

Computational Fluid Dynamics (CFD) test studies fall into two categories; 2-D simulations and 3-D simulations.

5.1 2-D simulations

Both Euler and viscous studies were conducted on the scramjet forebody, inlet, and isolator sections. 2-D Euler flow studies were conducted using the air vehicles unstructured solver (AVUS) [11]. AVUS is a three-dimensional finite volume unstructured-grid Euler/Navier-Stokes flow solver. 2-D isolator viscous simulations were conducted using an in-house computational scheme, the integral differential scheme (IDS) [12]. The following contour plots (**Figures 17–19** and **24–29**) represent the solution of the AVUS software, whose units are in the SI. Whereas the contour plots shown in **Figure 20** depict the solution of the IDS. The IDS is built on the premise of reducing numerical and modeling errors. As such, the IDS

Figure 17.
AVUS Euler results velocity distribution.

DOI: http://dx.doi.org/10.5772/intechopen.85697

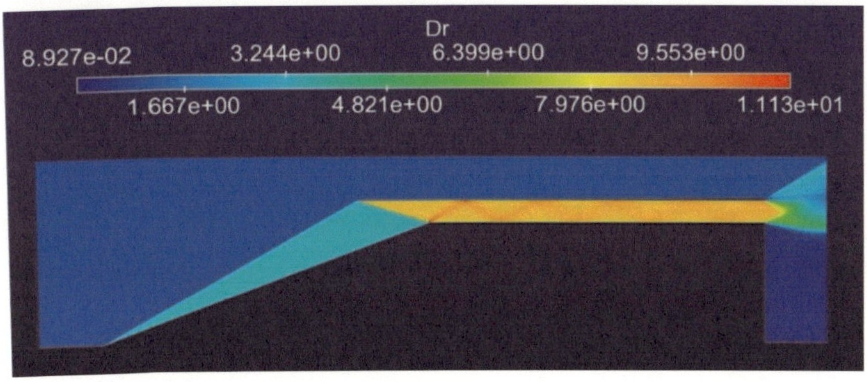

Figure 18.
AVUS Euler results density contours.

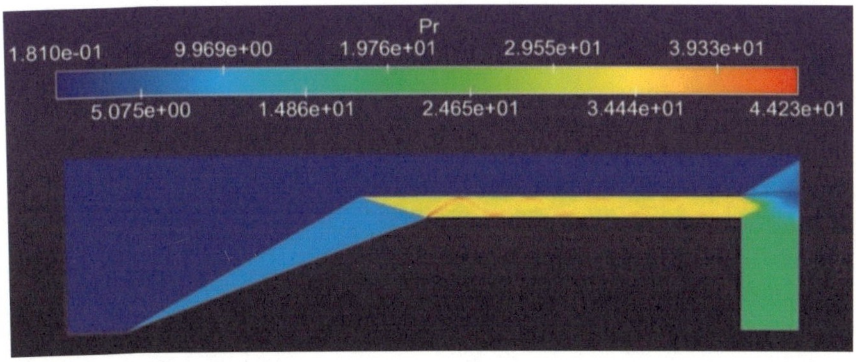

Figure 19.
AVUS Euler results pressure contours.

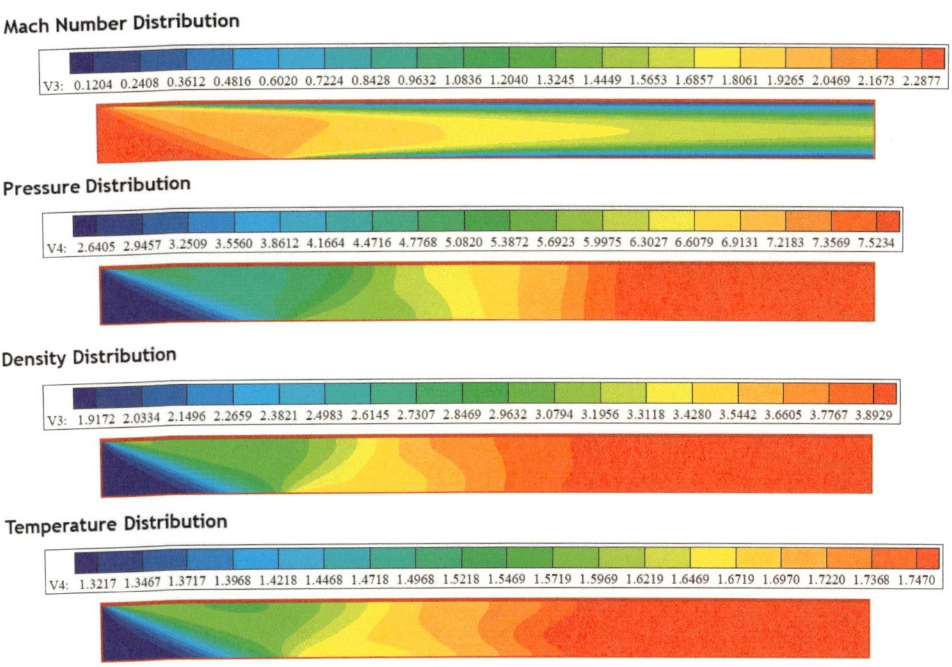

Figure 20.
Results from isolator 2-D IDS simulation.

Figure 21.
3-D stream tube in computational domain.

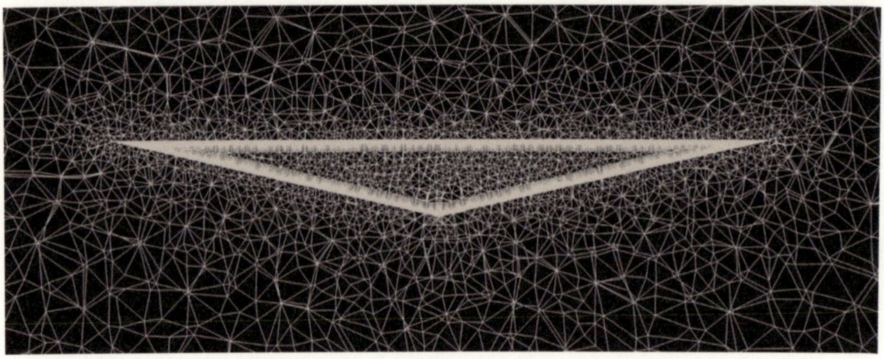

Figure 22.
3-D stream tube cross-section (isolator exit) with clustering unstructured grids.

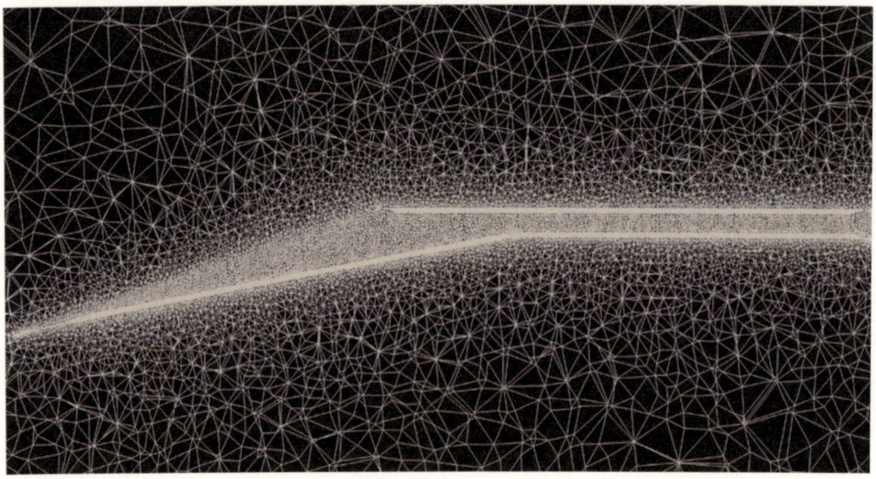

Figure 23.
3-D stream tube centerline with clustering unstructured grids.

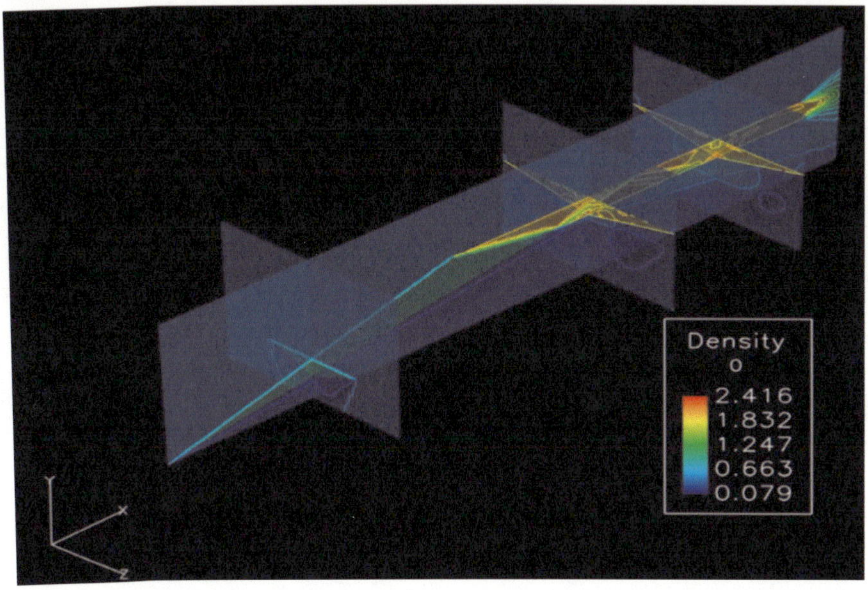

Figure 24.
Forebody-inlet-isolator validation study with 2-D slices.

Figure 25.
Centerline 2-D Mach number contours.

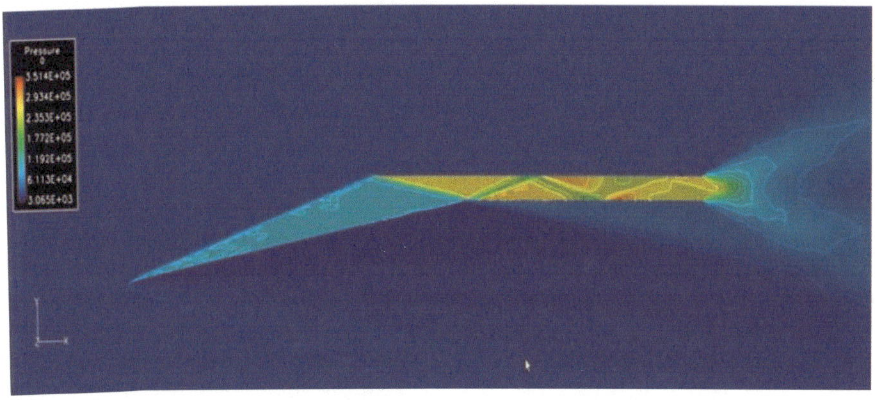

Figure 26.
Centerline 2-D pressure contours.

Figure 27.
Centerline z-component velocity contours.

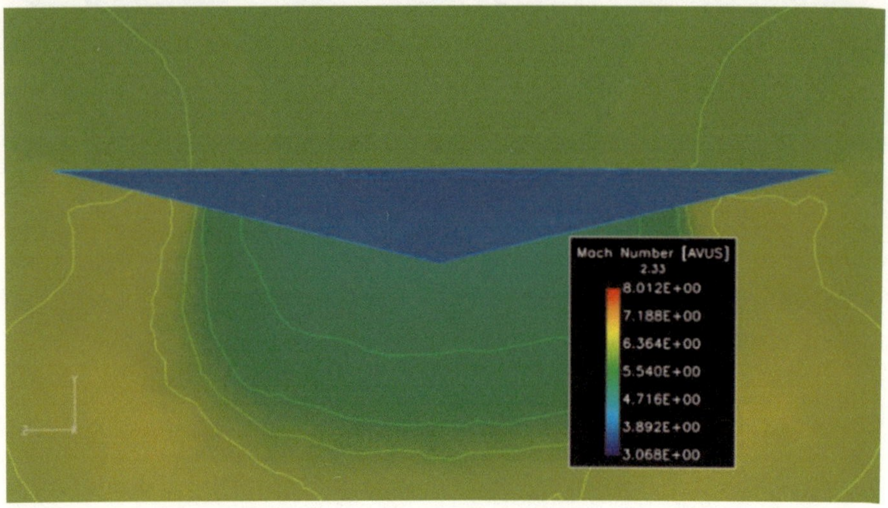

Figure 28.
Mach contours at the isolator exit.

implements the dimensionless form of the Navier Stokes equations, and therefore it reduces the round-off error.

5.1.1 2-D Euler simulations using AVUS

The four-point star configuration, **Figure 14**, was selected as the test case. The scramjet forebody-inlet-isolator model was exposed to a Mach 5 hypersonic freestream flow-field at a zero angle of attack. **Figures 17–19** presents the 2-D Euler simulation results along the centerline of the four point star configuration. On examining these figures, the following observations are made. **Figure 17** presents velocity distribution data for the geometry, where it is observed that the behavior of the flow imitates the conceptual flow-field presented in **Figure 4**. That is, freestream flow is first processed by the 2-D oblique shock, travels parallel to the wedge surface, is processed again by the reflected shock and travels parallel to the isolator duct walls. **Figures 18** and **19** presents the density and pressure flow-field

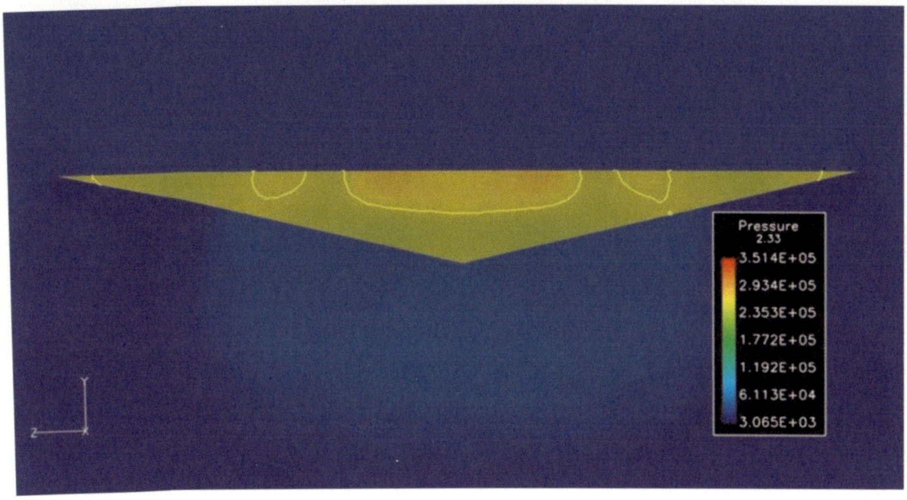

Figure 29.
Pressure contours at the isolator exit.

distributions. Once more we observe the organized nature of the 2-D flow, which is supported by the constant property in the respective zones. The development of the shock train within the isolator duct is also captured in these figures.

5.1.2 2-D isolator viscous simulations

A 2-D viscous simulation was conducted on the isolator section using the integral differential scheme (IDS), currently under development at North Carolina Agricultural & Technical State University. At the heart of the IDS numerical scheme is the unique combination of both the differential and integral forms of the Navier-Stokes equations (NSE). The differential form of the NSE is used for explicit time marching, whereas integral form of the NSE is used to evaluate the spatial fluxes. The IDS scheme has the ability to capture the complex physics associated with fluid flows. It does this by using a 'method of consistent averages' (MCA) procedure which ensures the continuity of the numerical flux quantities. The objective of this initial simulation was to observe the flow behavior. Further details on the physics and computational numerical scheme associated with the IDS can be found in [12]. **Figure 20** presents the flow-field distribution. Flow-field properties presented in **Figure 20** include Mach number distribution, pressure distribution, density distribution, and temperature distribution. Examination of theses flow-field properties supports the fact that the flow-field is behaving in a manner as it was designed to.

5.2 3-D simulations

3-D computational simulations were also conducted on the scramjet forebody, inlet, and isolator sections. Computational tools used were Fluent and AVUS. In the case of the 3-D Euler computational simulation, a single 3-D stream tube, **Figure 13**, was exposed to a Mach 6 flow-field. The simulation was first conducted using Fluent, where the process is summarized by **Figures 21–23**. Examining **Figures 22** and **23** demonstrates the use of unstructured grids with clustering in key areas for the analysis. The 3-D simulation required 6.7 million elements, 1,165,267 nodes, and 14.75 GB memory. To aid with visualization, 2-D slices, such as those seen in **Figure 24**, were extracted for analysis.

A similar process was implemented with AVUS and 2-D slices of flow-field data are extracted and presented in **Figures 25–29**. On examining these 2-D slices of 3-D data, it is observed that the stream tube is processing the flow in an organized consistent manner that is aligned with its design. Of note are **Figures 27–29**. **Figure 27** examines the z-component of the velocity, and indicated that there is very little cross-flow. Arguable, this is an Euler analysis, however it is worth pointing out that the stream tube 2-D design process holds. This is further supported by **Figures 28** and **29** which presents data on the Mach and pressure distribution at the isolator exit.

6. Combustor diffuser nozzle sections

A dual mode scramjet configuration, as presented in **Figure 30**, in addition to having a fore-section consisting of a forebody, inlet, isolator sections; also has an aft-section consisting of a combustor, diffuser and nozzle sections. Whilst the focus of this chapter has been on the design of the fore-section, a brief discussion on the design of the aft-section and its integration is warranted for the sake of completion in the design of the dual mode scramjet.

Design of the aft-section focuses on four fundamental design sections; a transition section, a combustor section, a diffuser section and a nozzle section, **Figures 31** and **32**. The transition section, as implied in the name, is designed to prepare the

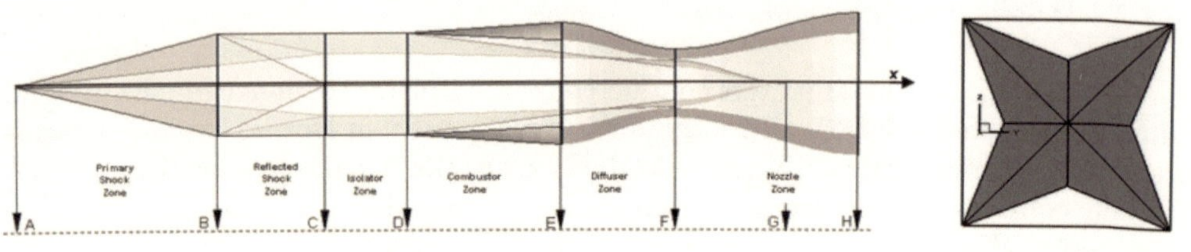

Figure 30.
A dual mode ramjet-to-scramjet concept [1].

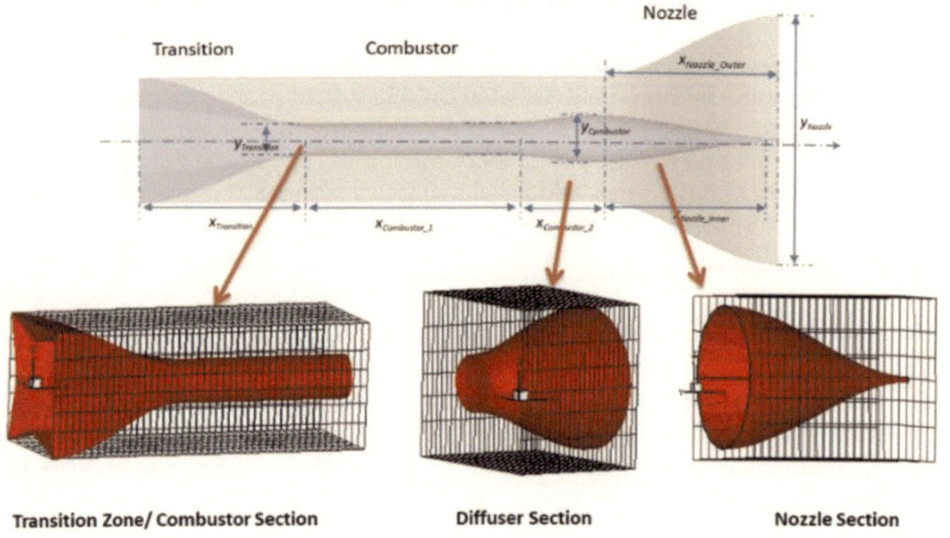

Figure 31.
2D-3D geometric construction from prescribes isolator cross-section and aerodynamic inputs [1].

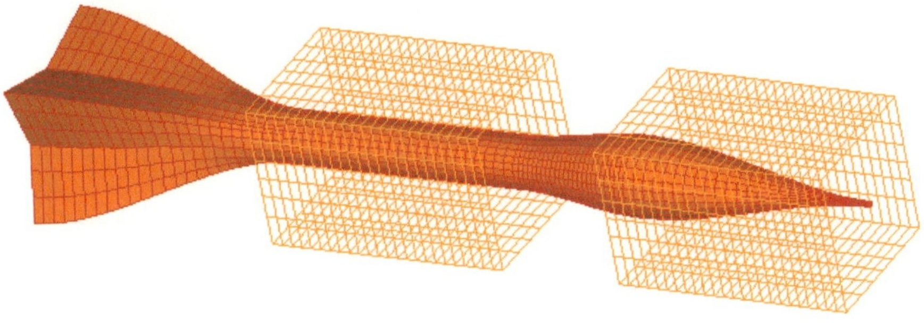

Figure 32.
Illustration of the transition-combustor-nozzle element.

flow before it enters the combustor. This section takes the flow leaving the isolator duct and guides it towards combustor. The primary design goal is to ensure that the flow entering the combustor is as organized as possible. The combustion section is where fuel is added, mixed and burned. The diffuser section is used to help control the combustion process as the scramjet operates across its dual mode, that is, switching from ramjet mode to scramjet mode. The nozzle section is used to accelerate the exhaust gases as the flow leaves the dual mode scramjet.

7. Scramjet flow-path

A completed scramjet flow-path can now be obtained with the assembly of both the forebody-inlet-isolator and combustor-nozzle sections. Two samples are presented in **Figures 33** and **34**. **Figure 33** presents a scramjet that has a square combustor configuration and **Figure 34**, a circular combustor configuration. Referring back to **Figure 2(c)**, one can observe that a variety of geometries can be

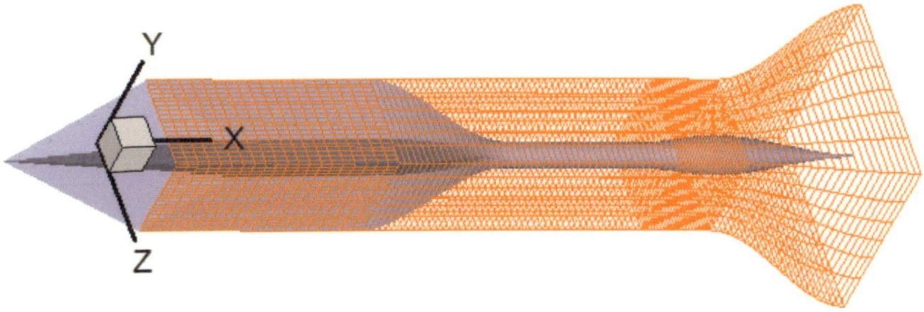

Figure 33.
4-pts scramjet with square combustor C-sections.

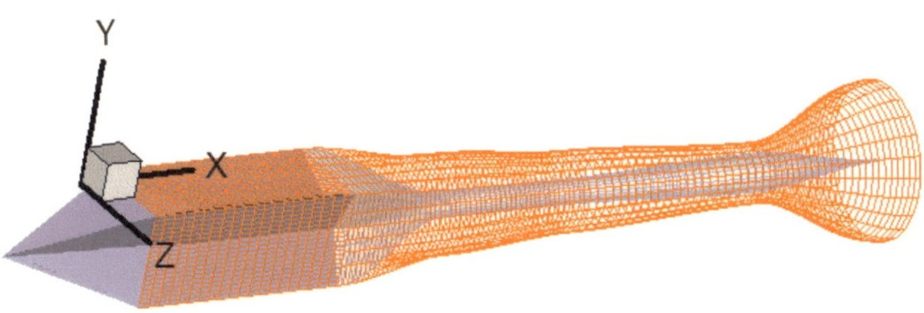

Figure 34.
4-pts scramjet with circular combustor C-sections.

generated by manipulating the design points A through H, and the design variables, x_3–x_{11}, in any combination.

8. Conclusion

In this chapter we explored an inverse design approach used in designing scramjet configurations. The forebody, inlet and isolator sections formed the core focus of the chapter. Ideal oblique 2-D shock relations along with Billig's isolator relations were first used to generate a centerline 2-D geometry. Streamline marching techniques coupled with Nonweiler's caret waverider theory were used to geometrically construct 3-D stream tubes. These 3-D stream tubes were later used in the construction of various star-shaped forebody-inlet-isolator sections. Initial 2-D Euler and 2-D viscous studies were performed on the forebody-inlet-isolator sections and the results presented. Initial 3-D Euler studies were also conducted on a single 3-D stream tube. All results presented demonstrated the uniform nature of the flow-field within the stream tube, supporting the inverse design approach. A 3-D viscous analysis of the 3-D stream tube is yet to be undertaken.

Acknowledgements

Special appreciation is extended to Dr. Isaiah Blankson of NASA Glenn Research Center for his continued guidance. The authors will also like to acknowledge the contributions made by the graduate students who have contributed to this on-going research effort. These include, Thomas Lawrence, Honest F. Mrema, Jovan Brown, Jamil Grant, and Nastassja Dasque.

Author details

Mookesh Dhanasar[1], Frederick Ferguson[1] and Julio Mendez[2*]

1 North Carolina Agricultural and Technical State University, Greensboro, North Carolina, United States of America

2 Corrdesa LLC, Tyrone, Georgia, United States of America

*Address all correspondence to: jcmendez@aggies.ncat.edu

References

[1] Dhanasar M. Development of A Benchmark for the Design and Analysis of A Tip-To-Tail Ramjet-Scramjet Propulsion System [dissertation]. Greensboro: North Carolina Agricultural and Technical State University; 2009

[2] Ferguson F, Dhanasar M, Williams R, Blankson I, Kankam D. Supersonic and hypersonic slender air-breathing configurations derived from 2D Flow-fields. In: 46th AIAA Aerospace Sciences Meeting and Exhibit; 7–10 January 2008; Reno, NV. DOI: 10.2514/6.2008-163

[3] Anderson JD. Fundamentals of Aerodynamics. 3d ed. New York: McGraw-Hill; 2001

[4] Anderson JD. Hypersonic and High Temperature Gas Dynamics. American Institute of Aeronautics and Astronautics. Virginia, USA: McGraw-Hill; 1989

[5] Anderson JD. Modern Compressible Flow: With Historical Perspective. 3rd ed. New York, USA: McGraw-Hill; 2004

[6] Heiser WH, Pratt DT. Hypersonic airbreathing propulsion. AIAA Education Series. Virginia, USA: American Institute of Aeronautics and Astronautics; 1994. ISBN: 1-56347-035-7

[7] Curran ET, Murthy SNB. Scramjet Propulsion. Vol. 189. Virginia, USA: AIAA; 2000. ISBN: 1-56347-322-4

[8] Billig FS. Research on supersonic combustion. Journal of Propulsion and Power. 1993;**9**(4):499-514

[9] Waltrup PJ, Billig FS. Prediction of recompression wall pressure distributions in scramjet engines. Journal of Spacecraft and Rockets. 1973; **10**(9):620-622

[10] Nonweiler TRF. Aerodynamic problems of manned space vehicles. Journal of the Royal Aeronautic Society. 1959;**63**:521-528

[11] Air Force Research Laboratory, CFD Research Branch, Wright-Patterson AFB, OH, Air Vehicles Unstructured Solver (AVUS)

[12] Ferguson F, Mendez J, Dodoo-Amoo D. Evaluating the hypersonic leading-edge phenomena at high Reynolds and Mach numbers. In: Recent Trends in Computational Science and Engineering. Rijeka, Croatia: IntechOpen; 2017

Chapter 5

The Busemann Air Intake for Hypersonic Speeds

Sannu Mölder

Abstract

A review and summary is presented of hypersonic air intake technology highlighting design objectives, basic flows, airframe integration, flowpath modification and intake flow startability. Taylor-Maccoll equations and Busemann flow are presented as the basis for constructing modular Busemann intakes. Wavecatching (streamline tracing), morphing and foreshortening are presented to show that (a) wavecatching is a useful technique to create modular startable intakes; (b) morphing is useful in integrating the intake shape with other geometric requirements of the airframe and combustor; and (c) foreshortening leads to minor gains in intake performance but large weight savings. A novel, *strong shock* method is presented, which uses strong-shock boundary conditions for designing spontaneously startable, modular Busemann intakes of high performance. This allows pre-determination of Busemann intake startability; offering great simplicity in the search for flowpath surface shapes that yield startable intakes with high compression, high efficiency and supersonic exit flows. Busemann flow contains unique fluid mechanical features: (a) a flow passage from a uniform, high Mach number flow, to another uniform, lower Mach number flow; (b) internal, convergent flow with an inflected surface; (c) conical flow where high gradients are near the center line and milder gradients are at the walls; (d) an axisymmetric and conically symmetric centered compression fan; (e) a free-standing conical shock, bounding irrotational flow. These are unique and fortuitous virtues, being significant in making the Busemann streamtube and its flow characteristics a suitable basis for designing high performance air intakes for hypersonic airbreathing engines.

Keywords: hypersonic air intake, Busemann design performance

1. Introduction to hypersonic air intake technology

There is a need for transporting man, machines, materials and munitions through Earth's atmosphere at high speed. Engines that propel fast airplanes are either rockets, or engines of airbreathing type. Turbojets, ramjets and scramjets (supersonic combustion ramjets) are types of airbreathing engines for propelling airplanes in the sensible atmosphere. The practical airbreathing engine for hypersonic speeds (above 5000 km/h) is the scramjet. A schematic is shown in **Figure 1**. The key components of the scramjet engine are the intake, the combustor and the nozzle. Mission studies have shown that a scramjet-propelled vehicle can provide a 2-hour travel time to most places on Earth or it can aid in the task of boosting vehicles to Earth orbit.

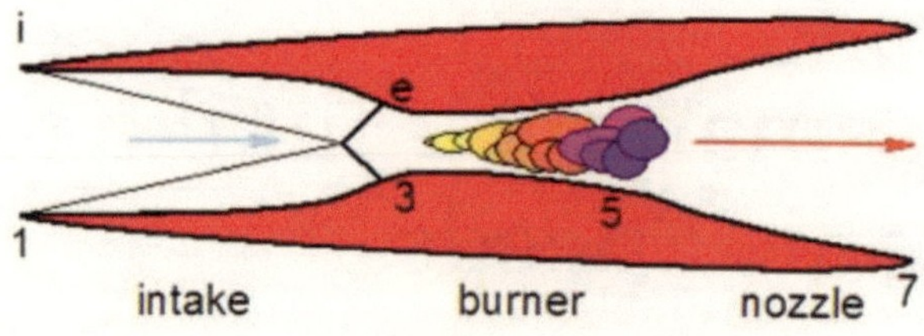

Figure 1.
The scramjet and its three components.

The air intake is one of the three essential components of supersonic and hypersonic airbreathing engines. It conditions the approaching freestream airflow for the combustor and the nozzle, compressing the airflow for best performance of the downstream components. For best overall engine performance, it must do so with minimal losses. The intake's performance can be thought of in terms of what the intake does and how well it is doing it-the *capability* and the *efficiency*. Capability can be quantified by such variables as the ratio of entry-to-exit Mach number, ratio of flow areas (contraction) or ratio of pressures (compression). Efficiency can be measured by the total pressure recovery or the entropy rise from entry to exit. These considerations of "what" and "how well" are governed by the First and Second Laws of Thermodynamics. Both capability and efficiency are highly dependent on the geometry of the intake surface. This raises additional design concerns about the intake's length and weight as well as exit flow inclination and uniformity, and the startability of the intake flow. These factors have a major effect on the intake's performance and on its design.

1.1 Intake design objectives

The hypersonic intake decelerates and compresses the freestream airflow as it flows into the combustion chamber. In doing this, (a) the intake flow must start and achieve the required decrease in Mach number with a high thermodynamic efficiency; (b) the compressed airflow, flowing into the combustor, should be uniform and stable; (c) the intake should operate efficiently and stably over the entire flight envelope bounded by flight Mach number, altitude and angle of attack; (d) the intake structure should be as light as possible; (e) drag of external surfaces and aerodynamic heating loads should be minimal; (f) for useful engine operation, the intake duct shape and flight conditions must be such that the intake airflow is predictable, properly conditioned (uniform in some sense) and aligned with the combustor walls as it flows into the combustor. These requirements for intake design are often contradictory, making it difficult to attain optimum individual operating conditions simultaneously. For example, in a fixed-geometry intake, flow starting and the need for substantial Mach number reduction pose a very serious geometric flowpath contraction contradiction that leads to an onerous design task.

Thermodynamic cycle calculations of high Mach number airbreathing engines, such as scramjets, have shown that the engines should have air intakes that contract and compress the flow by factors of 6–10 and 10–20 respectively and that this contraction and compression should be done with minimal loss of total pressure. Aside from high contraction and compression, the attainment of efficient intake performance is critically dependent on the freestream Mach number and the lateral and stream-wise contours of the intake surface, both being factors in determining the character of the flow in the intake and its performance.

Since all these design concerns are intake shape-dependent, it is most convenient and reasonable to start an iterative type intake design procedure with the selection of an intake shape that is known to produce a geometrically simple, compressive flow. Flow on a plane inclined wedge, Prandtl-Meyer flow and flow over a circular cone, as well as their combinations, have been used extensively as starting points for supersonic intakes because their aerodynamic characteristics are simple and easily predictable analytically. These "textbook" flows are usually adaptable to physical variations in geometry where shape change may be required for optimised performance over a range of Mach numbers and to ensure intake flow startability. In selecting such simple and easily predictable flows and their streamlines, for intake applications, we search for geometric streamlines that join a uniform and parallel freestream entry flow to an equally uniform and parallel exit flow. For most intakes the exit flow direction should be the same as that of the freestream. The flow compression in the intake should be longitudinally distributed so as to be isentropic at the high Mach number, upstream end of the intake. Minimal shock losses are obtained when compression through shock waves occurs at the lower Mach number, downstream end. So as to minimize viscous losses, all surfaces should contribute usefully to the compression task by individually supporting positive pressure gradients. The desirable qualities should not deteriorate significantly at off-design conditions of flight Mach number, altitude or angle of attack.

1.2 Simple planar and axial flows

Scramjet engine thermodynamic cycle calculations and combustor performance place a requirement on the hypersonic air intake to reduce the freestream Mach number by a factor of about three and to do so with a total pressure recovery of at least 0.5. These design targets can be met by employing combinations of simple inviscid flows that are assembled to form the intake flowpath. The simple flows can be based on either planarly symmetric (planar) or axially symmetric (axial) supersonic "text-book" flows. In planar flows, flow properties are the same in parallel geometric planes. In axial flows, flow properties are invariant in planes around a common axis. Because of planar or axial symmetry, the number of independent spatial geometric variables, needed to specify the flow, is reduced from three to two—a great simplification for design and analysis. These simple flows also possess radial symmetry in that there is no variation of flow properties along flat planes (planar flow) or cones (conical Taylor-Maccoll flow). Use of simple flows with flat plate and conical symmetry allows the number of spatial variables that are required to specify and describe the flow, to be further reduced by one, so that only one independent geometric variable remains—a further simplification for intake design and analysis. Examples of such commonly used simple planar flows are the flow behind a flat oblique shock and Prandtl-Meyer flow. Commonly used simple axial flows are the flow over a cone and the Busemann flow. Simple flows and their combinations do not carry shocks that are curved in the flow plane; this keeps the intake flows irrotational and uniform. An important part of intake design consists of combining and connecting the simple flows to yield the desired intake performance. The other part consists of using selected streamline sheets of these flows to form desired flowpath shapes—a technique called *wavecatching*.

The focus in this paper is on the use of *axial*, internal flow elements (basic flows) [1–3], rather than planar flow elements, to construct intake flow paths. The axisymmetric intake attains most of its compression by flow convergence rather than flow turning or shock deflection. The converging flow is isentropic, it is similar to sink flow and it causes a Mach number decrease which leads directly to weak terminal shocks waves and efficient intakes with high capability. The internal converging flowfield is the most important feature of an axial flow intake.

Problems of viscous losses and flow starting are eased by use of wavecatcher technology, providing leading edge truncation and sweep and by the fact that high adverse pressure gradients occur in the inviscid core flow rather than in the wall surface boundary layers. Hypersonic intakes that utilize axisymmetric compressive basic flows with specified entrance and exit shapes have received attention because of their high performance (capability and efficiency) and analytical simplicity [4–24].

A preferred geometry for a scramjet combustor is a duct with a circular cross section because of its superior ability to withstand both heat and pressure loads. Frictional losses are also at a minimum for such a duct since a cylinder has the smallest surface area for a given cross-sectional area. This leads to a cylindrical (axially symmetric) geometry as being desirable also for the intake that is attached to the front of the combustor duct. The same circular exit geometry for the intake is demanded by a gas turbine engine, in this case because the axial compressor face is circular. Towards these ends, it is pertinent to study an axisymmetric flow and it is entirely fortuitous that axisymmetric, conical, Taylor-Maccoll flow provides a streamtube shape [1, 2] that satisfies the above intake design requirements, both geometric/structural as well as aerodynamic [3]. In recognition of Adolph Busemann's pioneering work [1] on such streamtube shapes, they are called Busemann flows and Busemann intakes. References [1–18, 24] all concern Busemann flow.

1.3 Intake flow processes and inward/outward flows

The reduction of Mach number, in the various basic flows, is accomplished by one or more fluid mechanical mechanism: (a) compressive flow *turning*; (b) flow *convergence* with area contraction and compression in a converging passage and (c) flow *deflection* through an oblique shock. Flow turning and contraction are isentropic processes leading to no loss in efficiency. Flow deflection through an oblique shock entails an entropy increase—a loss in intake efficiency. If shocks are needed to deflect or re-direct the flow then they should be as weak as possible, occurring at the lowest possible Mach number (e.g., Busemann shock). Planar flow turning by Prandtl-Meyer-type flow requires much turning to accomplish a significant Mach number reduction, so that, after P-M turning, strong shocks are required to re-direct the flow back to the freestream direction for the combustor. On the other hand, isentropic Mach number reduction by area contraction leads to a rapid streamwise Mach number reduction when the flow is axial. In such flows, Busemann flow being typical, there is comparatively little flow turning towards the center line, the compression being accomplished by area contraction and, as a result, there is no need for much deflection (re-turning) by a shock at the exit. Also, since there is considerable Mach number reduction in the converging flow, the terminal shock faces a reduced Mach number. This weaker terminal shock minimizes efficiency losses. The axial flow intakes derive their high efficiency from the axial convergence, being only little degraded by flow deflection through the terminal shock. The axial Busemann intakes have been mistakenly labeled as "inward turning" even when part of their converging flow is turning outward, away from the axis. We suggest dropping the "inward turning inlet" terminology in favor of "axial flow intake" or "converging flow intake," because their fundamental and characterizing distinction is axial convergence. It is precisely the lack of much "inward turning" that leads to the high performance of Busemann intakes. It would be better to use the flow-related and meaningful concepts of *turning, convergence and deflection* to characterize intake flow types in general. Isentropic turning, as in

P-M flow, may have to be used where variable geometry demands the use of planar flow. The resulting flow turning, away from the flight direction, has to be compensated by lossy oblique shock deflections. On the other hand, flow convergence, such as occurs in sink flow, is an effective mechanism because it is isentropic and involves no flow turning. Deflection occurs through an oblique shock; it is non-isentropic and it should be used only when there is no other possibility of orienting the flow. It should not be used to reduce the Mach number. A practical flow, such as Busemann [1], incorporates all three of these aerodynamic mechanisms as they interactively contribute to intake performance.

The three modes of compression are illustrated in the Prandtl-Meyer intake, the Oswatitch intake and the Busemann intake (**Figure 2**). The Prandtl-Meyer intake obtains performance by isentropic *turning* through the compression fan, followed by *deflection* through the oblique shock; there is no *convergence*. The Oswatitch intake has flow *divergence* and *turning* followed by *deflection* through a shock. The Busemann intake has *turning* and *convergence* followed by shock *deflection*. Three intake models were designed to reduce the Mach number from 8.33 to 4.8 with a static pressure ratio of 26.8. All three intakes were tested in a gun tunnel [11] at Mach 8.33 and it was found that, for the same amount of contraction, the inviscid total pressure recoveries of the Busemann, Oswatitch and Prandtl-Meyer intakes were 0.983, 0.763 and 0.763. Experimental total pressure recoveries were 0.484, 0.485 and 0.240. The reason for the differences stems from the fact that the surface area and consequently the viscous losses, were greatest for the Prandtl-Meyer intake. Sidewalls, needed to contain the planar Prandtl-Meyer flow, did not preserve the intake's efficiency but contributed to the surface area and viscous losses. The lack of an extensive leading edge and attendant viscous flow contributed to the efficiency of the Oswatitch intake. These results illustrate the superiority of axial over planar basic flows where it is the Mach number reduction, achieved by convergence, that leads to the high performance of the Busemann intake.

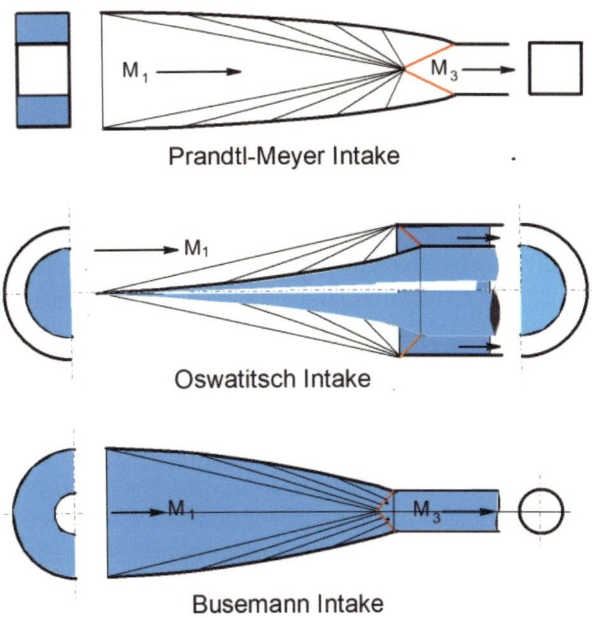

Figure 2.
Schematics of three intakes tested in a gun tunnel at Mach 8.33 [11].

1.4 Streamline tracing: wavecatching, morphing

The design technique of replacing known streamline sheets garnered from simple flows, by solid surfaces to generate aerodynamic shapes, known as "streamline tracing" has been applied to wing-body shapes [20]. The resulting airplane shapes are named "waveriders." The objective in waverider design has been to generate airplane-like shapes that produce high ratios of airplane lift-to-drag force. The same streamline tracing technique, in this case called "wavecatching," is applied to intake flowpath design [4, 7, 13, 14, 21–26] to generate intake surfaces. The objective, in this case, is to generate intake surfaces that capture, support and contain internal flows that have a high performance as supersonic/hypersonic air intakes. Both waverider and wavecatcher applications rest on the fluid-mechanical principle of replacing impervious streamline sheets by solid wall surfaces. In both applications the design starts with the selection of a freestream capture area cross-sectional shape. The shape is projected, as a closed trace, onto the leading shock wave of a prescribed simple flow. In the case of wavecatchers, the trace becomes the leading edge of the intake and the shock wave or Mach wave covers the leading edge of the intake at the design condition. All the mass flow passing through the trace is captured into the intake. The streamtube extending downstream from every point of the closed trace forms the shape of the intake's flowpath. By selecting a suitable shape for the entry flow trace, much design flexibility is available in integrating the resulting engine flowpath with the airframe shape and the intake's exit flow shape to suit the combustor. Pre-selecting the simple flow determines the internal flow as well as the intake performance. The technique is equally applicable to planar and axial flows as well as to non-symmetric flows. The basic wavecatcher technique, when applied to symmetric flows, produces geometrically similar flowpath cross sections. Various methods of morphing can be applied to gradually change the flowpath shape if the entry and exit cross sections are not to be geometrically similar [7, 11, 26]. Two very important extra virtues of the wavecatcher design method are that it produces flow paths with swept leading edges, much like a sugar scoop where, at the design Mach number, the leading shock is everywhere attached to the leading edge; there is no flow spillage. However, at off-design conditions, such a swept leading edge does permit overboard mass spillage during intake flow starting, making otherwise non-startable intakes startable. Experimental results on wavecatcher intake shapes, based on Busemann flow, were presented in [4]. Using streamline tracing methodology, based on the flow in a straight conical duct, the notion of selecting portions of the axisymmetric versions of internal flow was used also in [21–23, 27]. The significant virtues of wavecatcher intakes has been utilized in many subsequent intake studies [10, 13, 21–23, 28–30]. The technique of streamline tracing, to produce modular flowpaths of arbitrary cross sectional shape (wavecatching), results in the following attractive aspects: (a) the total mass flow is divided between individual modules, reducing the mass flow demand of test facilities, both wind tunnels and flight test, (b) thrust vector control is easier to implement with flow in individual modules, (c) highly swept module leading edges make mass flow spillage possible for intake flow starting, (d) module freestream capture shapes are easily integrated with airframe shapes, (e) modules can be raised off the airframe surface so as not to ingest the fore-body boundary layer, (f) properly designed modules are self-startable.

Wavecatching and morphing techniques for modular Busemann intakes will be discussed in Section 7; module startability in Section 8.

1.5 Intake starting/unstarting

For spontaneous ignition and supersonic combustion, the Mach number at combustor entry should be about one-third of the flight Mach number. In a flight Mach number range 4–25, the intake cross-sectional area must decrease by a factor of 5–20. Such a highly convergent duct can support two distinctly different flow configurations at any given supersonic flight Mach number. One flow type produces a bow shock in front of the intake that diverts much flow overboard and, in this case, the intake flow is subsonic with unacceptably low performance. This is termed "subcritical" or "unstarted" flow. The other possible flow configuration has no bow shock, no overboard spillage and is supersonic throughout. This "supercritical," or "started" flow, is required for efficient scramjet engine operation. Attainment of supercritical flow in high contraction ratio intakes present a problem in that the intake flow will not assume the started flow state spontaneously under steady flight conditions. Starting requires that the near-normal bow shock, in front of the unstarted intake, moves downstream into the intake to be "swallowed" and that a stable hypersonic/supersonic flow is established throughout the converging portion of the intake. Spontaneous starting will not occur in intakes whose exit-to-entry area ratio is below 0.6. Unfortunately, startable intakes with exit-to-entry area ratios at or above this value do not produce enough compression to be useful as scramjet intakes. Methods of intake flow starting must be found and implemented for high contraction intake flowpaths. Intake starting is not open to design compromises; it is a critical, non-negotiable requirement that presents challenges and places severe conditions on intake design. Since startability is determined, to a limited extent, by flight Mach number, there is some design flexibility in choosing the start Mach number. Various methods of promoting intake flow starting have been explored in [9, 11, 18, 25, 27, 31–37]. Section 8 presents an analytical approach to the design of spontaneously starting, modular Busemann intakes.

Unstarting of started flow is also a concern in that flight at an extreme angle of attack or at combustor overpressure conditions can cause the intake to regurgitate a stopping shock and the intake flow to revert from a started condition to a condition of unstart. Such an event must be prevented since it is followed by a catastrophic loss of thrust. A review of research progress on detection and control of unstart mechanisms of hypersonic inlets is described in [27, 34].

2. Taylor-Maccoll equation(s) and Busemann flow

Busemann [1] described an axisymmetric, conical flow that starts in the uniform freestream, compresses and contracts isentropically and passes through a conical shockwave to become uniform and parallel to the freestream flow. Courant and Friedrichs [2] make a brief reference to Busemann flow, suggesting its use as an air intake. Molder and Szpiro [3] used the Taylor-Maccoll equations to calculate the inviscid Busemann flow and present a capability/efficiency performance map for the flow as a hypersonic air intake. Experiments, at Mach 8.33, on a full Busemann intake and on modular, wavecatcher surfaces, based on Busemann flow were conducted by Mölder and Romeskie [4] and by Jacobsen et al. [25]. VanWie and Molder [12] suggested applications of the Busemann intake to hypersonic flight vehicles.

The Busemann intake shape is analytically defined by only two numerical parameters [3]. This has made it easily "transportable" and led to its proposed use as a benchmark standard for internal flow CFD verification [38], and a basis for more general studies of intake flows as well as experiments for such issues as flow starting [18, 25, 27, 31–37], viscous effects [15, 39], truncation [16, 18], drag measurement [11], wavecatcher configurations [13], leading edge blunting [10] and cross section morphing [7]. Viscous effects and truncation and stunting are found in [15, 18, 39]. Experimental results for full and modular Busemann intakes are found in [11, 13, 18, 21, 23, 25, 33, 37, 40]. A four-module Busemann-based intake on a scramjet engine was launched at Mach 5, from a large ballistic gun [41].

The high performance [24] and analytical simplicity of the Busemann intake has made it a subject for some 60 publications.

2.1 Description of Busemann intake flow

The basic Busemann intake surface is axisymmetric (**Figure 3**). It is a converging duct with its axis aligned with the freestream. When started, it captures freestream flow (M_1) in a circular cross section. Since there is no flow deflection at the leading edge there is a zero-strength conical Mach wave (io) from the leading edge at the freestream Mach angle. The flow then starts turning towards the axis, so that flow area decreases and pressure increases in the flow and along the surface (icfs). A maximum turning angle (inflection point) is reached at (f). Turning from (i) to (f) has made the flow convergent so as to compress by convergence. While still convergent and inclined towards the axis, the flow (fs) starts turning away from the axis, passing through a conical shock (os) where it is deflected to become uniform and parallel to the axis at the exit of the intake. This "turning away" of the pre-shock flow lessens the flow deflection requirement of the terminal shock, leading directly to an increase in efficiency. It is this efficiency increase and the convergence in (icfsoi) that contribute directly to the superior performance of the Busemann intake. Flow in the region (icfsoi) is isentropic and irrotational. In the region (icoi) the compression waves from the surface (ic) converge to the focus (o). The compression waves from the rest of the surface (cfs) are incident on the terminal shock. Aside from axial symmetry, this flow is also conically symmetric so that there is a focal point (o) on the axis, from which rays can be drawn, in any direction, such that the flow conditions on any ray are constant. Except for the leading ray (io), the rays are not Mach waves. Axial symmetry makes the rays to be generators of cones so that the flow conditions on circular cones are constant and the conditions are functions of only the conical angle θ (**Figure 3**). All the streamlines are geometrically self-similar, with shapes that are scalable with distance from the origin. Thus, only one streamline, $r = f(\theta)$, needs to be calculated to define the intake surface. Disappearance of the radial dimension (r) as an independent variable, in conically symmetric flow, permits the depiction of all flow conditions on the

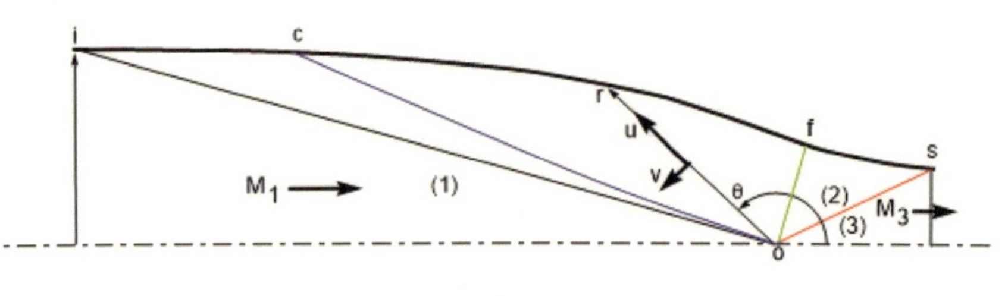

Figure 3.
Busemann intake contour is icfs. io *is a freestream Mach cone.* os *is a conical shock. Uniform entry flow at (1). Uniform exit flow at (3). Supersonic, isentropic, axially and conically symmetric flow from (1) to (2). Flow crosses oblique conical shock from (2) to (3), C-characteristics in* ico *focus at* o. *C-characteristics from* cfs *are incident on the shock along* os. *All streamlines have an inflection point on the cone* fo. *Spherical-polar coordinate system* (r, θ) *is centered at* o *with corresponding radial and angular Mach number components* u *and* v.

single remaining spatial variable—the conical angle, θ. This offers great simplicity in flow analysis where a wide variety of intake surfaces is available for selection of surface shapes that yield both a high compression and a high efficiency for the intake. Furthermore, the presence conical flow means that all shocks, facing conical flow are also conical and therefore of constant strength, at any angular position. The flows are not only uniform but also irrotational—generally a desirable feature for flow that leaves the intake to enter a combustion chamber. These features of conical flow and, in particular, Busemann flow, which is by nature an internal, compressive flow, make the basic Busemann streamline shape an attractive candidate for constructing an air intake for a hypersonic flight vehicle's engine.

2.2 Flow symmetry: coordinate axis: flow direction

Flow which is both axially and conically symmetric is best described in spherical polar coordinates (r, θ, ϕ) where r is distance measured radially out from the origin, θ is the angle measured counterclockwise from the downstream direction and ϕ is the circumferential coordinate around the axis of symmetry (**Figure 4**). For Busemann flow the origin is at the apex of the conical shock, on the center line of symmetry (xx). The flow velocity components in the radial and angular directions are designated as U and V. Drawing similar triangles along the streamline, in **Figure 4** gives the streamline equation:

$$dr/d\theta = rU/V = ru/v \tag{1}$$

Busemann flow, and axisymmetric conical flow are governed by the Taylor-Maccoll equation, the same equation that governs the supersonic flow over an axisymmetric cone at zero angle of attack. The original Taylor-Maccoll equation is a non-linear, second order total differential equation with the spherical polar angle, θ, as independent variable and the radial flow velocity, U, as dependent variable [42, 43].

$$\frac{\gamma-1}{2}\left[1-U^2-\left(\frac{dU}{d\theta}\right)^2\right]\left[2U+\frac{dU}{d\theta}\cot\theta+\frac{d^2U}{d\theta^2}\right]-\frac{dU}{d\theta}\left[U\frac{dU}{d\theta}+\frac{dU}{d\theta}\left(\frac{d^2U}{d\theta^2}\right)\right]=0 \tag{2}$$

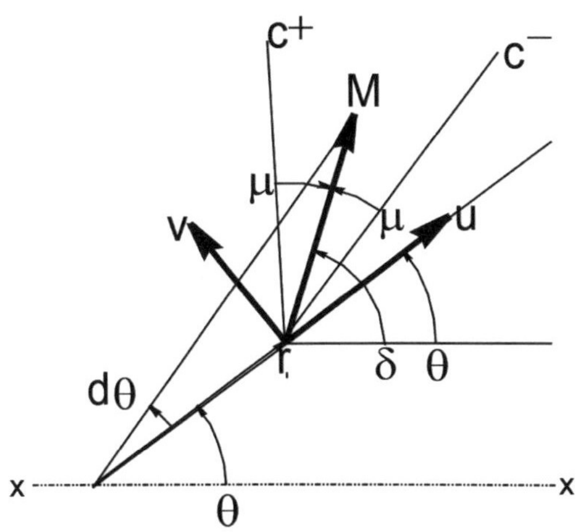

Figure 4.
The coordinates (r, θ)*; the Mach number* (M) *and its radial and angular components* u *and* v.

This is the model equation that governs steady, axisymmetric, conical flow of a perfect gas. No explicit algebraic solution has been found, nor are there any numerical schemes for solution of the second-order Eq. (2) as given above. However, the equation can be converted to two first order Eqs. (3) and (4), at the price of acquiring the additional dependent variable, V. But the two equations are now amenable to standard numerical solution methods. Most of these solutions have been done with boundary conditions applicable to flow over an axisymmetric cone [42, 43].

2.3 The first-order Taylor-Maccoll equations

The first-order versions of Eq. (1) are the momentum equations, in spherical polar coordinates, in the r and θ directions [44]:

$$dV/d\theta = -U + \frac{a^2(U + V\cot\theta)}{V^2 - a^2} \tag{3}$$

$$dU/d\theta = V \tag{4}$$

where a is the speed of sound that can be written in terms of the velocities and the total conditions through the energy equation. The second of these equations is also the irrotationality condition, implying that conical flows are necessarily irrotational. Explicit reference to the speed of sound and total conditions can be circumvented if the equations are recast so as to have the radial and angular Mach number components (u, v) as dependent variables in place of the corresponding velocity components (U, V). The boundary conditions, when expressed as Mach number components at the up- and down-stream sides of conical shocks, are then applicable directly to the solution of the equations. Also, total conditions, which have no influence on the Mach number solution, do not have to be invoked.

2.4 Mach number components (u, v) as dependent variables

The Taylor-Maccoll (T-M) Eqs. (3) and (4) have been recast in terms of the radial and angular Mach numbers u and v, where $u = U/a$ and $v = V/a$ and a is the local sound speed:

$$\frac{du}{d\theta} = v + \frac{\gamma - 1}{2} uv \frac{u + v\cot\theta}{v^2 - 1} \tag{5}$$

$$\frac{dv}{d\theta} = -u + \left(1 + \frac{\gamma - 1}{2} v^2\right) \frac{u + v\cot\theta}{v^2 - 1} \tag{6}$$

These two equations seem more complicated than their parents (3) and (4). However, it will be shown that the use of Mach number components u and v leads to meaningful and useful physical interpretations from Eqs. (5) and (6). Also, the sound speed has been eliminated as a variable.

In terms of Mach number components, the streamline Eq. (1) is,

$$dr/d\theta = ru/v \tag{7}$$

and the flow Mach number is,

$$M = \sqrt{u^2 + v^2} \tag{8}$$

Having the T-M equations in this form reveals their singular nature at $v = \pm 1$ where the singularity is caused by the $(v^2 - 1)$-term in the denominators above.[1] The term $u + v \cot\theta$, appearing in both numerators, is the component of Mach number normal to the axis. This component is zero for the freestream flow, so that, at the entrance, the Taylor-Maccoll equations take on a 0/0 type singularity and it turns out that $(u + v \cot\theta)/(v^2 - 1)$ has a finite value at the freestream entrance of the Busemann intake.

As a result of using the Mach number variables u and v, the absence of any explicit reference to total conditions, as well as the sound speed, leads to a more straightforward application of the boundary conditions. A standard, fourth-order Runge-Kutta scheme [45] has been used to integrate the Mach number components, u and v form of Eqs. (5) and (6) and $r = f(\theta)$, from Eq. (7). The solutions are identical, to eight decimal places, to similar solutions of (5) and (6) in the velocity variables. Eqs. (5) and (6) govern and describe the flow in a Busemann intake and Eq. (7) gives the streamline/surface shape.

3. Solution of the Taylor-Maccoll equations

Eqs. (5) and (6) are simultaneous, first-order, total differential equations that can be solved by standard methods, such as in Ralston and Wilf [45], for the two Mach numbers u and v in terms of θ. The Mach number M is then found from $M = \sqrt{u^2 + v^2}$ and other thermodynamic values follow from isentropic relations. The shape of the intake surface can also be integrated within the integration routine to give r in terms of θ, $r = f(\theta)$ so that the Cartesian coordinates of the axisymmetric Busemann surface shape are found from $x = r\cos\theta$ and $y = r\sin\theta$.

3.1 Boundary conditions at shock and freestream

Integration of Eqs. (5) and (6) requires the starting values u and v at the value of $\theta = \theta_2$ in front of the shock. A convenient and aerodynamically significant approach is to select the Mach number in front of the shock M_2 and the aerodynamic shock angle θ_{23} as the starting variables. The flow deflection through the shock, δ_{23}, is found from the equation relating Mach number, shock angle and flow deflection [28]:

$$\tan\delta_{23} = \frac{2\cot\theta_{23}\left(M_2^2 \sin^2\theta_{23} - 1\right)}{2M_2^2(\gamma + 1 - 2\sin^2\theta_{23})} \tag{9}$$

The angular location of the shock, which is the starting value for the variable of integration, θ, is then:

$$\theta_2 = \theta_{23} - \delta_{23} \tag{10}$$

This ensures that the flow behind the shock is parallel to the axis, which is the most common requirement of flow entering a combustor. The starting values for the radial and circumferential Mach numbers are then:

$$u_2 = M_2 \cos\theta_{23} \tag{11}$$

[1] Such singularities are discussed in [29, 45, 46]. Their appearance, in any given flow, should be taken as a warning that whatever symmetry assumption(s) have been made may not hold in the physical airflow.

$$v_2 = -M_2 \sin \theta_{23} \tag{12}$$

The radial variable, r, becomes dependent on u and v and the starting value r_2 at the shock. The value of r_2, the shock's length, is arbitrary at this stage. It determines the scale size of the streamline and its utility becomes relevant when considering morphing and wavecatching in Sections 1.3 and 7. Note that, prior to integration of Eqs. (5) and (6), and calculation of the intake surface shape, we could calculate the intake's efficiency, using the total pressure ratio as measure,

$$p_{t3}/p_{t2} = \left[\frac{(\gamma+1)k^2}{(\gamma-1)k^2+2}\right]^{\frac{\gamma}{\gamma-1}} \left[\frac{\gamma+1}{2\gamma k^2-\gamma+1}\right]^{\frac{1}{\gamma-1}} \tag{13}$$

and the capability from the exit Mach number,

$$M_3^2 = \frac{(\gamma+1)^2 M_2^2 k^2 - 4\left(k^2-1\right)\left(\gamma k^2+1\right)}{\left[2\gamma k^2-(\gamma-1)\right]\left[(\gamma-1)k^2+2\right]} \tag{14}$$

where $k^2 = M_2^2 \sin^2 \theta_{23}$ is the square of the shock-normal Mach number component. In fact, we could prescribe a desired efficiency, p_{t3}/p_{t2}, and calculate k^2 from Eq. (13); also prescribe the downstream Mach number M_3 and calculate M_2 by inverting Eq. (14). Then $\theta_{23} = \sin^{-1}(k/M_2)$, $u_2 = M_2 \cos \theta_{23}$ and $v_2 = -M_2 \sin \theta_{23}$. After this, θ_2 and δ_{23} are found as above and the integration performed, on increasing θ, until $(u + v \cot \theta) \geq 0$. The ability to specify the downstream Mach number and an intake efficiency, before doing the integration, makes this approach particularly suitable for preliminary intake design selection. Note, however, that *all is not roses*, since the integration yields a freestream Mach number that may not be the desired one. An iteration, on the input conditions, p_{t3}/p_{t2} and M_3, or k^2 and M_2 has to be performed to arrive at the desired intake design Mach number. This inconvenience is the direct result of, and the price paid for, the convenience and simplicity achieved by imposing the flow to be conically symmetric and by imposing the outflow conditions. It turns out that, using the T-M equations, the flow curvature and gradients of pressure and Mach number can also be found at the shock wave before the complete integration is done (Sections 4.4–4.6).

Eqs. (5) and (6) are then numerically integrated from θ_2 *to* $\theta_1 = \pi - \mu_1$ in an upwind direction with an increasing θ. Since θ_1 is not known *a priori*, the integration is continued until the normal-to-the-axis (cross-stream) Mach number $(u \sin \theta + v \cos \theta)$ becomes zero or positive, indicating that the freestream has been reached. The calculated shape and Mach number contours of such an integration are shown in the top half of **Figure 5**.

Note the conical nature of the contours. The calculated Busemann shape is then used as input to a CFD code to predict the flow as shown in the lower half of

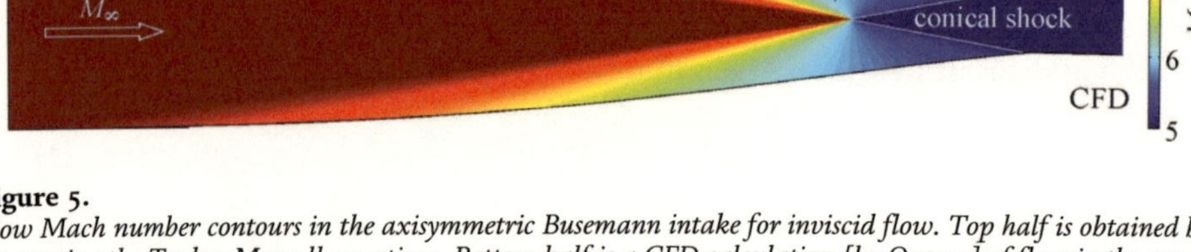

Figure 5.
Flow Mach number contours in the axisymmetric Busemann intake for inviscid flow. Top half is obtained by integrating the Taylor-Maccoll equations. Bottom half is a CFD calculation [by Ogawa] of flow in the same intake shape as the top half.

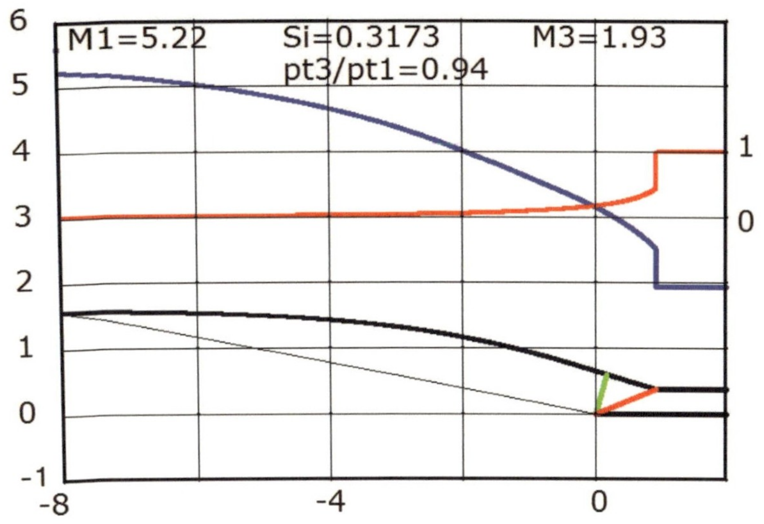

Figure 6.
Busemann intake contour (black curve) with conical shock (red) and cone of inflection points (green). Mach number distribution (blue). Pressure distribution, normalized with respect to exit pressure (red, on the right side ordinate), for an intake that reduces the Mach number from 5.22 to 1.93 with a total pressure recovery of 0.94.

Figure 5. Although the CFD code is not "told" anything about conicality, the conical nature of the flow is well represented by the CFD calculations. Both methods predict a uniform exit flow downstream of the conical shock (courtesy Dr. Ogawa). This is an illustration of the use of Busemann flow as a benchmark for verifying the application of a CFD code to internal flow. Graphical results of an integration of Eqs. 5–7 are shown in **Figure 6** for a Busemann intake that reduces the Mach number from 5.22 to 1.93 with a total pressure recovery of 0.94.

3.2 Singularity at entry

At the entry, Busemann flow joins to the freestream at a conical Mach wave. The Mach number normal to this wave, $v = -1$, which makes both Eqs. (5) and (6) have a zero in their denominators. At the conical Mach wave $u + v \cot\theta$ is also zero so that Eqs. (5) and (6) have a 0/0-type singularity. This makes it impossible to start the integration at a specific freestream Mach number so as to progress in a clockwise (downstream) direction towards the shock. An infinite number of streamlines are possible and unique boundary conditions cannot be specified at the freestream. The starting value of r_2 is arbitrary; it determines the scale size of the streamline and its utility becomes relevant when considering morphing and wavecatching in Sections 7.1 and 7.2.

4. Aerodynamic features of Busemann intake

This section describes some features of Busemann-type intake flow that are unique to axisymmetric conically symmetric flow. First, there is the geometric simplicitly that arises from the axial and conical symmetries. These symmetries require that conditions on a circle, which circumscribes the axis, are constant and conditions are constant also on any circular cone surface whose axis is aligned with the symmetry axis and whose apex is confocal with all other such cones.

Another very important feature is the fact that all solutions of the T-M equations, starting from an acute angled, conical shock, always end up at a straight and parallel freestream flow. Busemann flow would be useless, as the basis for an air intake, if this were not so. This fortuitous feature must be inherent in the T-M Eqs. (5) and (6). This property of the T-M equations holds whether the downstream flow is set to be uniform or not, as long as it is conical.

The downstream end of the Busemann flow has an inflection point where the surface turns away from the axis, towards being parallel with the exit flow. This lessens the flow deflection required from the terminal shock and also lessens the strength and loss produced by the terminal shock. This feature contributes directly to the high efficiency of the Busemann intake flow.

4.1 Free-standing conical shock

In a parallel, uniform, freestream a conical axisymmetric shock is produced by a conical body and the shock strength is proportional to the cone angle. In Busemann flow there is no solid cone, yet a conical shock is produced. This "free-standing" shock is possible because the flow in front of the shock is converging towards the center line and the center line behind the shock is acting as a zero-angle cone to force the flow into a parallel and uniform downstream direction. Such a free-standing conical shock, with uniform post-shock flow, is unique to Busemann flow.

Experiments were conducted, in a Mach 3 wind tunnel, at the Defence Research and Development Canada (Valcartier) laboratories to demonstrate the existence of the free-standing conical shock [40]. Since a full Busemann intake would not start spontaneously in the steady wind tunnel flow and, also, since the shock would be hidden from tunnel optics by a full Busemann duct, only an annular, leading edge portion of the Busemann duct was constructed and tested (**Figure 7b**). The tip of the conical shock, produced by the annulus, is in the region of influence of the annulus and that was sufficient to produce a freestanding conical shock at the center line that was in the field of view of the tunnel optics (**Figure 8**). Compression waves, from the annulus, converge to the center line and reflect as a conical shock as calculated by CFD in **Figure 7a**. No incident shock or Mach reflection is apparent. Calculated post-shock Mach number is 1.48, and pressure is 10.1 and temperature is 1.94 times their freestream values.

The yellow arrow points to the focal point where the converging compression fan and the free-standing conical shock meet. The analytically predicted Busemann flow and its features have been confirmed by both CFD and experiment. The approach presented here is the only method for establishing a centered axial compression followed by a conical shock at the center line in a steady flow. Flow properties inside the apex of the conical shock can be precisely set and the shape of

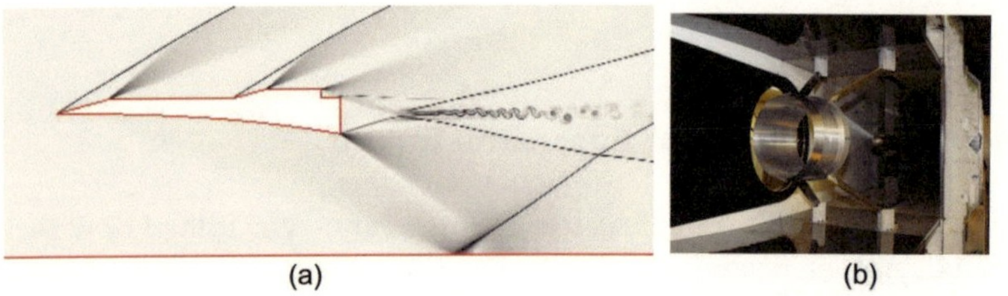

Figure 7.
(a) Freestanding conical shock at center line, produced by axisymmetric Busemann leading edge annulus in a Mach 3 freestream. CFD calculation by E.V. Timofeev. (b) Busemann leading edge annulus in Mach 3 wind tunnel at DRDC [40].

Figure 8.
Freestanding conical shock in Busemann flow at Mach 3 (DRDC). Blue arrow points to apex of conical shock.

the required Busemann annulus calculated so as to create a local high pressure hot spot for igniting a supersonic fuel/air mixture at a precise location.

4.2 Characteristics

Characteristics are two sets of intersecting lines in supersonic flow. The characteristics carry a physical significance in that they delineate the region of space that influences flow conditions at a particular point as well as the region of space that depends on the flow conditions at a point. The characteristic lines are selected such that, along these lines, the governing partial differential equations become total differential, finite difference equations, allowing numerical solutions of the flow-field [42, 46].

Alternatively, once a supersonic flow-field has been calculated by some non-characteristic methods, the characteristic lines can be calculated and superimposed and inferences about influences, causes and effects can be drawn. The α and β or C+ and C− characteristics are inclined at $\pm\mu$ to the local streamlines where $\mu = \sin^{-1}(1/M)$ (**Figure 9**). In polar coordinates the α and β characteristics' shapes are determined by integrating,

$$\left(\frac{dr}{d\theta}\right)_{\alpha,\beta} = r\cot(\delta - \theta \pm \mu) \tag{15}$$

where the plus sign is for the α characteristic and minus is for the β characteristic. For x-y plotting one can integrate the α-characteristics directly:

$$\begin{aligned}(dx/d\theta)_\alpha &= r\cos(\delta+\mu)/\cos(\pi/2-\delta-\mu)\\(dy/d\theta)_\alpha &= r\sin(\delta+\mu)/\cos(\pi/2-\delta-\mu)\end{aligned} \tag{16}$$

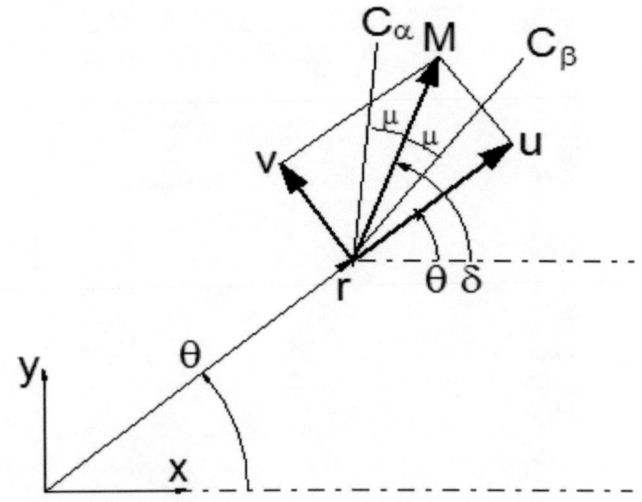

Figure 9.
Characteristics C_α *and* C_β.

and the β-characteristics by:

$$\begin{aligned}(dx/d\theta)_\beta &= r\cos(\delta-\mu)/\cos(\pi/2-\delta+\mu)\\(dy/d\theta)_\beta &= r\sin(\delta-\mu)/\cos(\pi/2-\delta+\mu)\end{aligned} \quad (17)$$

Integration of the characteristics is easily performed inside the routine for integrating the T-M equations. This method was used to superimpose characteristics on the T-M solution above. Resulting characteristic lines are shown in **Figure 10** for the same Mach 5.22 intake as in **Figure 6**.

4.3 Centered compression fan

The Taylor-Maccoll equations point to the existence of a confocal, conical, compression fan—the axisymmetric analogue to a Prandtl-Meyer fan. Such a fan of coalescing characteristics, preceding a free-standing conical shock, is shown to exist experimentally (**Figure 8**), as well as by CFD calculations (**Figures 7a** and **10**).

The characteristics mesh in **Figure 11** is a schematic overlay on the Busemann flow. The α-characteristics (not shown) all start from the freestream Mach cone and proceed away from the axis to intercept either the surface streamline or the front surface of the shock. The blue and red β-characteristics start at the surface and proceed towards the axis. The first of the β-characteristics is the freestream (1) Mach cone itself, having an inclination μ_1 at the axis. At the shock (2) the remaining characteristics have an inclination $\delta_2 + \mu_2$, different from μ_1. **Figure 11** is a

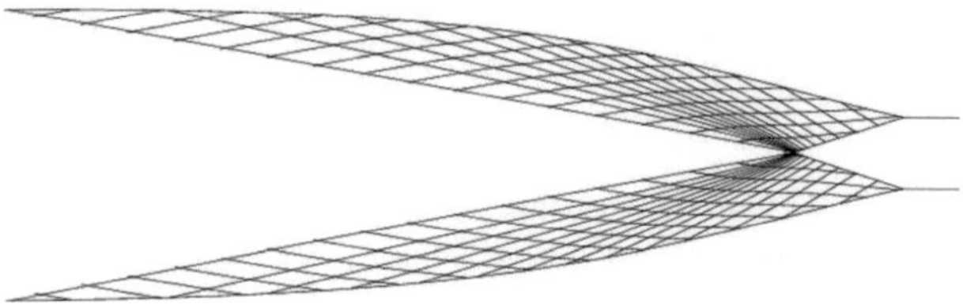

Figure 10.
α and β characteristics network for the Mach 5.22 Busemann intake. α characteristics are outbound from the center line and β characteristics are inbound. Note convergence of β characteristics at center line.

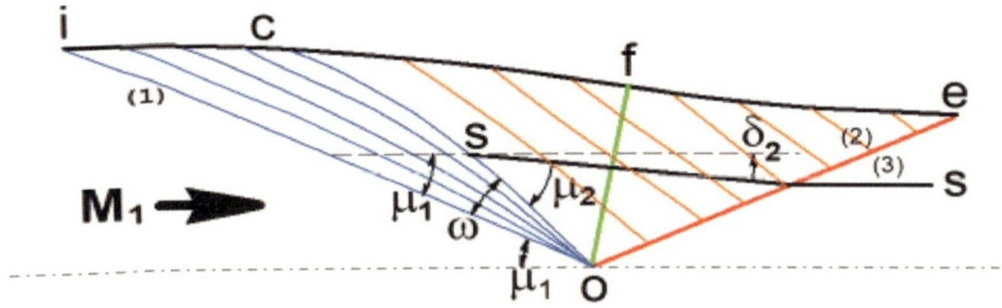

Figure 11.
Schematic of characteristics in Busemann flow. Centered compression fan (ioc)*. Shock-impinging characteristics* (cfeo)*. Subscript* (1) *refers to freestream conditions,* (2) *refers to pre-shock conditions. Angles shown are for conditions at* O.

zoomed-in schematic of the conditions at (o) showing a Busemann streamline (icfe) that joins a freestream (1) to the conical shock (oe) at (2). A streamline (ss) passes through the shock. The characteristic (os) and its projection to (c) is the last of the centered β-characteristics (blue) and it is also the first of the β-characteristics (red) that start at the surface and proceed towards the axis but intercepts the shock (oe).

An examination of the inclinations of the characteristics shows that the angular width of the centered compression fan, $\omega = \mu_2 + |\delta_2| - \mu_1$ must be >0, because $\mu_2 > \mu_1$ (since $M_2 < M_1$), so that $\omega > 0$ and the fan must exist. The angular region ω is populated by β characteristics that fan out from (o) to the Busemann streamline along (ic). The fan of β-characteristics contained in (oci) is a centered, axisymmetric compression fan, analogous to the Prandtl-Meyer fan in planar flow. The shape of the last centered characteristic (oc) and the location of (c) can be calculated during the integration of the intake flow when the variable of integration, θ, reaches the value $\pi - (|\mu_2| + |\delta_2|)$. The red β-characteristics from the surface (cfe) all intercept the shock (oe) where a very small, near-apex segment of the shock, is determined by a relatively long length of the Busemann intake surface (ic). The rest of the shock shape is determined by the characteristics from the surface (cfe). This large surface-to-shock length ratio suggests that the leading edge shape is unimportant in determining the overall shock shape. However, a long leading edge surface length contributes to boundary layer growth and viscous losses, providing a reason and an incentive to truncate the leading edge so as to minimize the sum of leading edge shock and boundary layer losses on a practical intake surface. The results presented here give an indication of the extent (ic) to which the conical shock is influenced by a shortening of the intake surface (truncation). A study of viscous/inviscid efficiency loss tradeoffs by truncation or stunting should take direction from the location of point (c). Previous treatment of the centered confocal compression fan or the free-standing conical shock has not been found in the open literature.[2]

4.4 Surface curvature, $D = d\delta/ds$; inflection point

An equation for the curvature of the T-M streamline is derived to show that the streamline can have points of zero curvature—inflection points. The Busemann streamline has two points of zero curvature where one of these points has significance in the starting of a Busemann-type intake. The conical surface containing all inflection points in a typical Busemann flow is shown in green in **Figure 11** where

[2] An analog of this flow exists in planar flow where the region (ioc) is a Prandtl-Meyer compression fan, the region (cof) is then uniform, the shock (ok) is plane and the flow aft of the shock is again uniform. That is the flow topography in the Prandtl-Meyer intake.

the portion of the surface (icf) is turning towards the axis and the portion (fk) is turning away.

To derive an expression for the curvature of the T-M streamline we use the defining equation of the streamline,

$$dr/d\theta = ru/v \tag{18}$$

where u and v are the radial and angular components of Mach number as used in the T-M equations. Taking another θ-derivative of (7) gives,

$$\frac{d^2r}{d\theta^2} = -r\frac{u}{v^2}\frac{dv}{d\theta} + \frac{r}{v}\frac{du}{d\theta} + \frac{ru^2}{v^2} \tag{19}$$

In polar coordinates, (r, θ) the curvature of a planar curve is [28, p. 34],

$$D \equiv \left(\frac{\partial \delta}{\partial s}\right) = \frac{r^2 + 2\left(\frac{dr}{d\theta}\right)^2 - r\frac{d^2r}{d\theta^2}}{\left(r^2 + \left(\frac{dr}{d\theta}\right)^2\right)^{3/2}} \tag{20}$$

Eliminating the derivatives of r with Eqs. (7) and (18) gives,

$$D = \frac{r^2 + 2(ru/v)^2 + r^2\frac{u}{v^2}\frac{dv}{d\theta} - \frac{r^2}{v}\frac{du}{d\theta} - (ru/v)^2}{(r^2 + r^2u^2/v^2)^{3/2}} \tag{21}$$

In this expression the derivatives $dv/d\theta$ and $du/d\theta$ are given by the Taylor-Maccoll Eqs. (5) and (6) so that the streamline curvature is,

$$D \equiv \frac{d\delta}{ds} = \frac{uv(u + v\cot\theta)}{r(v^2 - 1)(v^2 + u^2)^{3/2}} \tag{22}$$

This equation gives the curvature of the T-M streamline in terms of the polar coordinates, r and θ, and the radial and polar Mach number components, u and v. It is plotted as the black curve in **Figure 12**. A number of very interesting and important features, about the T-M streamline, become apparent from an examination of its curvature as given by Eq. (21):

1. D is inversely proportional to r so that when $r \to 0$ then $D \to \infty$. This means that streamlines near the origin of T-M flows are highly curved. This is a necessary condition for flow over a cone, where flow, near the tip and just aft of the conical shock, has to rapidly adjust to the inclination demanded by the cone surface, since the flow deflection produced by the conical shock is insufficient for the flow to be tangent to the cone surface. Similar highly curved streamlines are to be expected near the focal point of Busemann flow. Conical flow is not conically symmetric, i.e., independent of r, when it comes to *gradients* of its dependent variables, such as streamline curvature—the dependence being inversely proportional to r. This inverse dependence on r extends to other flow property gradients as well.

2. There is an asymptotic condition, ($D = 0$) in the T-M streamlines at $v = 0$. For flow over a cone, $v = 0$ at the cone surface. This confirms that the streamlines become asymptotic to the cone surface as they approach the surface. There is no $v = 0$ asymptotic condition in Busemann flow.

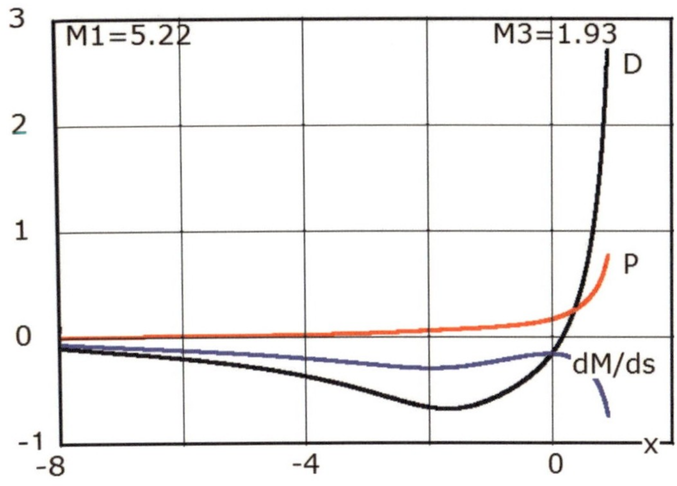

Figure 12.
Surface curvature (D)*, pressure gradient* (P) *and Mach number gradient* (dM/ds)*,* vs. *axial distance (x) in the isentropic part of the Busemann intake. x = 0 is at the apex of the conical shock. Highest values are reached at the corner where x = 0.98.*

3. When $u = 0$ then $D = 0$. This means that the streamline has a point of inflection at the place where the radial Mach number is zero. For flow over a cone the condition $u = 0$ never occurs, so the streamlines are curved monotonically positive. However, for Busemann flow there is a location, $\overline{\theta}$, where the streamline changes from being concave towards the axis (negative curvature) to being convex (positive curvature). The flow changes from turning inward, towards the axis to turning outward, away from the axis. At the inflected surface there is no turning, the flow is purely convergent. Numerical integrations of the T-M equations have shown that $\overline{\theta}$ always lies in the interval θ_2 to $\pi/2$ (first quadrant), somewhat upstream of the Busemann shock, as shown by the green line in **Figure 11**. Every Busemann streamline has an inflection point and, for each intake, these points form a unique conical surface. At this angular location the flow is everywhere normal to the green cone surface, whose half-angle is $\overline{\theta}$, and a conical normal shock can be placed coincident with the green cone since the Mach number is supersonic. This condition leads to an analysis for determining the startability of a wavecatcher Busemann intake according to the following argument: If the bow shock could be coaxed into taking up the inflection position by allowing enough mass spillage to occur between the shock and the inflection location and by restricting the downstream contraction to that allowable by the Kantrowitz criterion for flow starting, then the intake would start. The important variables in the Kantrowitz criterion are the "green" inflection cone surface area, the Mach number in front of the "green" cone and the exit area. These variables are available at the integration of Eqs. (5) and (6) at the streamline inflection angle, $\theta = \overline{\theta}$. If the contraction downstream of the conical normal shock surface does not lead to choking, then the shock moves downstream and the intake starts spontaneously. The starting event and its causes are critical in self-starting supersonic/hypersonic air intakes. It is a conical and axisymmetric example of the starting criterion posed by Kantrovitz for one-dimensional flow, embodying the same principle of flow choking downstream of a normal shock where, in this case, the normal shock is not flat but has a conical shape. Flow just downstream of the conical normal shock at the inflection point is inclined towards the axis. This ($r \rightarrow 0$)-type singularity is similar to the cone-

tip singularity described above; its existence, in the idealized form, has not seen confirmation by experiment or CFD. The cone of streamline inflections is a significant feature for assessing startability of wavecatcher Busemann intakes in Section 8.

4. There is a point of zero curvature also when $(u + v \cot\theta) = 0$. The quantity $(u + v \cot\theta)$ is the component of Mach number normal to the flow axis. For Busemann flow it is zero only where the Busemann flow joins the freestream. Thus, the leading edge of the Busemann flow has not only zero deflection but also zero curvature. Aerodynamically this means that the leading edge wave is neither compressive nor expansive but is a zero-strength Mach wave. The fact that the entering freestream flow is neither deflected nor curved by the Busemann leading edge means that the leading edge of a hypersonic air intake, based on Busemann flow, is ineffective in contributing to the intake's task of reducing the Mach number. This provides an incentive to foreshorten some length of the leading edge surface so as to decrease viscous losses, possibly without incurring serious inviscid flow losses. For M-flow [32] the potential appearance of the condition $(u + v \cot\theta) = 0$ is prevented by the appearance of the $(v \to \pm 1)$-singularity (described below) so that the post-shock flow never becomes parallel to the freestream. This is unfortunate from a practical viewpoint since it presents no possibility of grafting any of the flows that have a uniform upstream, such as cone or Busemann flows, to the downstream of M-flow. From a fundamental viewpoint it also presents an obstacle to the possibility of conical shock reflection at the center line of symmetry [31].

5. When the angular component of Mach number $v \to \pm 1$ then $D \to \infty$; the curvature becomes infinite and the streamline has a cusp or a corner. This indicates a *singularity* or a *limit line* at a corner. Neither cone nor Busemann flow exhibit such a limit line. However, it does occur in both M- and W-flows [32].

6. The quantity $(v^2 + u^2)^{3/2}$, appearing in the denominator of Eq. (21), is just M^3. It is always a positive quantity for all flows and has no drastic characterizing effect on D except to force streamlines to be less curved, to straighten out, at hypersonic speeds. Hypersonic intakes become long and slender.

4.5 Surface Mach number gradient in the flow direction, *dM/ds*

The streamline equation may be written,

$$\frac{ds}{d\theta} = \frac{r}{v}\sqrt{u^2 + v^2} \tag{23}$$

The flow Mach number, M, in terms of its radial and axial components, u, and v, is,

$$M = \sqrt{u^2 + v^2} \tag{24}$$

So that,

$$M\frac{dM}{d\theta} = M\frac{dM}{ds}\frac{ds}{d\theta} = u\frac{du}{d\theta} + v\frac{dv}{d\theta} \tag{25}$$

giving,

$$\frac{dM}{ds} = \frac{v}{r(u^2+v^2)}\left[u\frac{du}{d\theta} + v\frac{dv}{d\theta}\right] \tag{26}$$

where the derivative terms, in the square brackets, are given by the Taylor-Maccoll Eqs. (5) and (6), when multiplied by u and v, respectively.

$$u\frac{du}{d\theta} = uv + \frac{\gamma-1}{2}u^2v\frac{u+v\cot\theta}{v^2-1} \tag{27}$$

$$v\frac{dv}{d\theta} = -uv + v\left(1+\frac{\gamma-1}{2}v^2\right)\frac{u+v\cot\theta}{v^2-1} \tag{28}$$

so that,

$$\frac{dM}{ds} = \frac{v}{r(u^2+v^2)}\left[1+\frac{\gamma-1}{2}\left(u^2+v^2\right)\right]\frac{u+v\cot\theta}{v^2-1} \tag{29}$$

This is the Mach number gradient, expressed in terms of the coordinates (r, θ) and the corresponding Mach number components (u, v), where the Mach number component values come directly from the integration of the Taylor-Maccoll Eqs. (5) and (6). dM/ds is plotted in **Figure 12** (with s measured in the downstream direction).

4.6 Surface pressure gradient in the flow direction, $P = (dp/ds)/(\rho V^2)$

In the isentropic flow, from the freestream to the shock, the gradients of Mach number and pressure are related by [47],

$$\frac{dM}{M} = -\frac{1+\frac{(\gamma-1)}{2}M^2}{\gamma M^2}\frac{dp}{p}$$

$$\text{or}\quad \frac{dM}{ds} = -\left[1+\frac{(\gamma-1)}{2}M^2\right]MP$$

where $\frac{dM}{ds}$ is given by Eq. 26; so that the non-dimensional pressure gradient is,

$$P \equiv \frac{dp/ds}{\rho V^2} = \frac{-v^2}{r(u^2+v^2)^2}\frac{(u+v\cot\theta)}{(v^2-1)} \tag{30}$$

The pressure gradient is expressed in terms of the radial and azimuthal coordinates r and θ and the radial and angular Mach number components u and v. It is plotted in **Figure 12**. This permits the calculation of the surface pressure gradient from quantities obtained in the T-M calculation of Busemann flow.[3] It also means that the surface pressure gradients are known everywhere on the surface of the highly three-dimensional wavecatcher shapes where all the surface gradients are useful as inputs to boundary layer calculations. Towards this end it is noted that, for the flow just upstream of the corner, where the shock impinges, the u and v Mach numbers are given by u_2 and v_2 from Eqs. (11) and (12), so that the gradients immediately before the shock-boundary-layer interaction at the corner can be evaluated just from the prescribed initial conditions using Eqs. (5)–(7), before embarking on a calculation of Eqs. (5) and (6). This enables a selection of initial conditions that is based on considerations involving the shock losses as well as the shock-boundary-layer interaction

[3] These gradient equations are applicable to all types of Taylor-Maccoll flows [48].

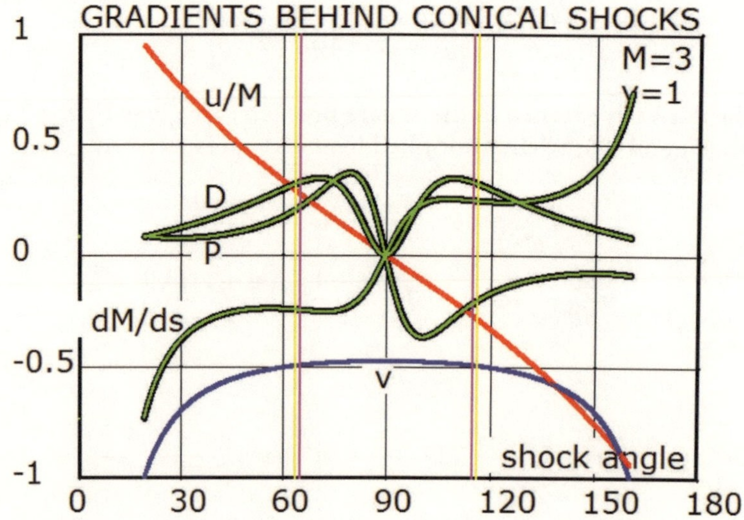

Figure 13.
Mach number components (u, v) *and gradients behind conical shock at Mach 3.*

effects. The analytical expressions for all three gradients have the radial coordinate r in their denominators. This requires the gradients to be the mildest on the intake surface and highest at the origin—a desirable condition for orderly wall boundary layer development on the intake surface.

4.7 Gradients at conical shockwaves

As a check on the various algebraic results we have plotted them against the acute (20–90°) and obtuse (90–160°) angles of conical shocks for Mach 3 in **Figure 13**. The left half of this figure is for acute shocks and the right half is for obtuse shocks, i.e., cone flow and M-flow. u (red) and v (blue) are the Mach number components behind the shock in the (r, θ)-directions. Black curves are for the various gradients from the T-M Eqs. (21), (26) and (27). The green curves are for the same gradients as calculated by Curved Shock Theory (CST) [48]. There is perfect agreement between gradients calculated from the T-M equations and those from CST. This is reassuring since the two methods are based on widely differing theoretical approaches.

5. Performance of Busemann flow as an air intake (inviscid flow)

An integration of the TM-Eqs. (5) and (6) from the initial conditions (10–12) is terminated when $(u + v \cot \theta) = 0$ at the free-stream where we discover the Mach number M_1. The results of many such calculations are shown in **Figure 14** where each complete Busemann intake calculation is represented by a dot. For each case, a value of M_2 is selected, in our case between 1 and 8, and k (the shock-normal component of M_2) is cycled from 1 to M_2. For each M_2 and k the total pressure ratio, Eq. (13), and M_3, Eq. (14), are calculated. Integration of the T-M equations then leads to the freestream at M_1 and a point is plotted on a graph of M_1 vs. M_3 with p_{t3}/p_{t1} as parameter, determining the point's color. Every point in this figure represents a Busemann intake calculation from the downstream shock to the freestream. This graph can be used to select a Busemann intake design based on the desired entry

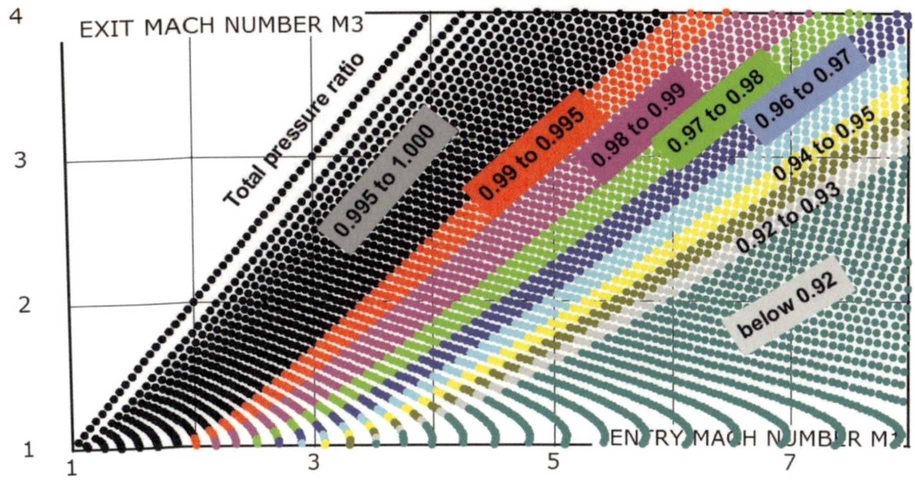

Figure 14.
Inviscid performance of Busemann intake.

and exit Mach numbers and the total pressure recovery. Any two of these parameters determine the third. For example, it is apparent from the graph that a Busemann intake that reduces the freestream Mach number from 7 to 3 does so with a total pressure recovery of 0.95. This graph represents both components of Busemann intake performance, the capability by M_1 and M_3 and the efficiency by p_{t3}/p_{t1}. Tradeoffs between these are workable with this diagram. As an example, a Busemann intake that reduces the Mach number by a factor of three does so with a total pressure recovery of about 0.90. A more refined and elaborate version of such a performance map is found in [3].

6. Boundary layer effects

High performance intakes have to have a very weak leading edge shock. Such a weak shock is inclined at near the Mach angle. This leads to the length-to-height ratio of the intake to be approximately M, the freestream Mach number. So that high performance intakes, including the Busemann intake, tend to become long and slender with large surface areas that have high shear near the leading edge, causing disproportionately high viscous losses. Surface length also leads to thick boundary layers at the exit with losses and the potential for major flow disruptions by boundary layer separation.

A comparison of inviscid and viscous flow in the Busemann intake is shown in **Figure 15** by Mach number contours. The blue, low Mach number boundary layer, appears in the viscous flow. The effect of the boundary layer has led directly to the presence of a shock from the leading edge and a noticeable change in the flow at the center line, a change of exit Mach number from 5.3 to 4.8 and a reduction in total pressure recovery from 0.97 to 0.43. The boundary layer has a significant effect on the inviscid flow even when it appears to stay attached.

Flow displacement by the boundary layer causes a conical shock to appear and focus to a point on the center line ahead of the Busemann flow focal point and a reflected, conical shock appears downstream that impinges on the surface ahead of the corner, **Figure 16**. To restore the inviscid flow topology and pressure distribution of the Busemann flow it is necessary to correct the surface shape of the intake

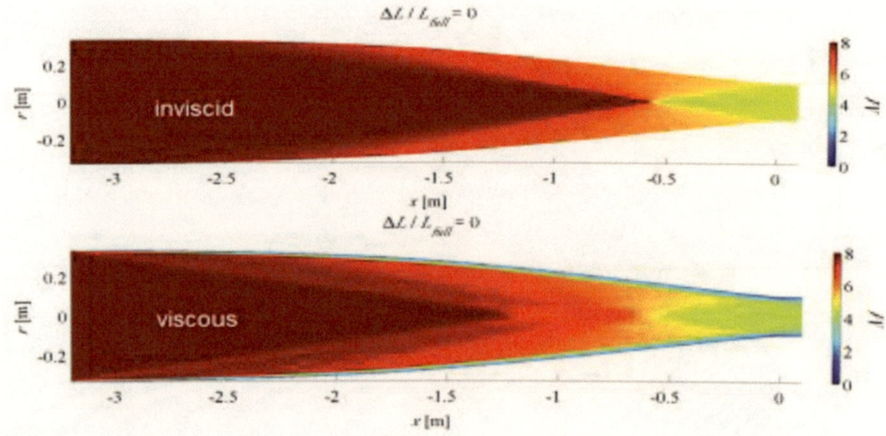

Figure 15.
Inviscid and viscous flow in the Mach 8 Busemann intake at 30 km flight altitude [20].

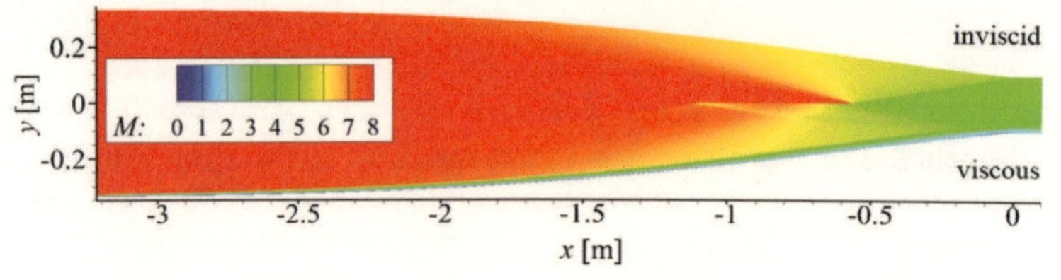

Figure 16.
Mach 8 Busemann intake flows, upper half without and lower half with boundary layer showing the effect of boundary layer on the wave structure of the inviscid flow in the unmodified Busemann intake.

by the boundary layer displacement thickness. The importance of viscous correction methodologies has attracted considerable attention and research efforts as accurate calculation of the boundary layer displacement thickness plays a pivotal role in intake performance assessment. Complex interactions of the shock waves and boundary layers developed on the curved surface of the Busemann intakes pose a challenge to accurate detection of the boundary layer edge. A viscous correction was applied [39, 49] to the full and truncated Busemann intakes by using the displacement thickness obtained through numerical integration of the CFD-generated boundary layer properties. Reasonable detection of the boundary layer edge was attained by examining the total enthalpy profile [50, 51]. Viscous correction is applied typically once only to produce the final geometry. However, the importance of repeating the process, with subsequent iterations, has been highlighted in [52] with the application of an updating procedure of the displacement thickness. The results of correcting for the boundary layer effect are shown in

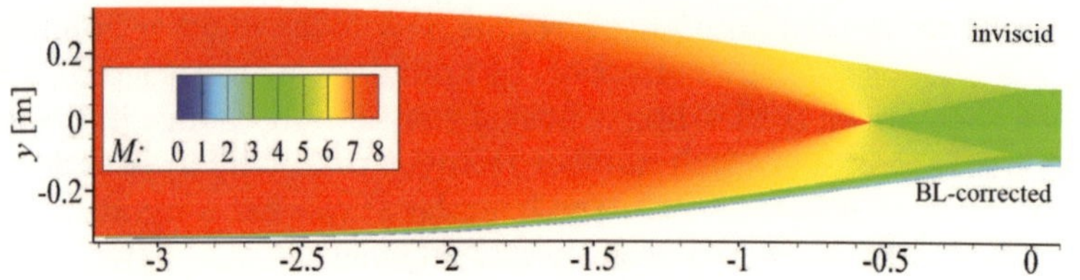

Figure 17.
Upper half contains inviscid Busemann flow. Lower half of flow is contained in a Busemann surface that has been corrected (enlarged) for boundary layer presence [17]. Note successful restoration of BL-corrected flow to be similar to inviscid flow.

Figure 17 where the bottom half of the figure shows the inviscid and viscous flow on a surface that has been obtained by enlarging the inviscid surface by the boundary layer displacement thickness. There is close resemblance between the inviscid and the boundary layer corrected inviscid flows in this figure. Both **Figures 16** and **17** were calculated by H. Ogawa.

7. Geometric modifications to Busemann flow: wavecatching, morphing, truncation, leading-edge blunting

The basic Busemann flow is contained in an axisymmetric streamtube of high contraction. As an intake, such a shape will not start at steady flow conditions. Also the axisymmetric shape may not conform well to the shape of the rest of the airplane surface nor the desired combustor entry and a need arises to modify its cross-sectional shape. Such modifications can be done while still retaining the basic Busemann flow characteristics by tracing the streamlines of the Busemann flow. This process depends on scaling and assembling adjacent, scaled streamlines into streamline sheets that form the wall surfaces of the intake module. The technique produces a wavecatcher intake module. In such constructions a chosen freestream capture cross-section shape becomes mirrored in a smaller, but geometrically similar, intake exit cross-section shape. If done properly, a wavecatcher module has a swept leading edge that captures the leading shock wave and mass flow at design conditions but permits flow spillage and promotes intake flow starting at design and off-design conditions. So a wavecatcher design gets away from an axisymmetric flowpath shape and it also leads to a startable intake as a separate outcome.

The wavecatcher intake shape, that integrates well with the airplane, may have an exit shape that is not necessarily the best shape for the combustor. The combustor shape is very likely wanted to be circular because it is to join to the contiguous combustor duct which is strongest and least aerodynamically lossy when it is circular. There is thus a need to deform the intake flow path gradually from the freestream entry to the exit; typically, from a segment of a circle to that of a full circle or possibly to an ellipse, (**Figure 18**). The method of doing this is called morphing [5, 7, 22, 23].

The Busemann intake has a large amount of surface immediately behind the leading edge. This surface carries a thin boundary layer and a high shear stress, contributing disproportionately to boundary layer losses. The question arises: Can boundary layer losses be decreased by foreshortening some of the surface aft of the leading edge? Realizing that truncation of the leading edge or stunting the intake will result in leading edge flow deflection and shock losses which counter gains achieved from decreased boundary layer losses.

Difficulties of cooling sharp leading edges lead to the adoption of leading edge blunting. Even a small amount of leading edge blunting can have a significant effect on the Busemann flow both in the boundary layer and in the inviscid stream [10].

7.1 Wavecatching (streamline tracing)

The objective of wavecatching is to generate intake flowpath surfaces different from the basic axisymmetric surface of the Busemann flow. The design starts with selecting the desired Busemann flow and calculating its streamline shape, $r = f(\theta)$, as in **Figure 3**. Wavecatcher intake surfaces are then generated from adjacent Busemann streamlines, $r = y(\phi)f(\theta)$ where r is a radial coordinate on the streamline, $y(\phi)$ is a scaling factor that varies smoothly from streamline to streamline and $f(\theta)$ is the shape

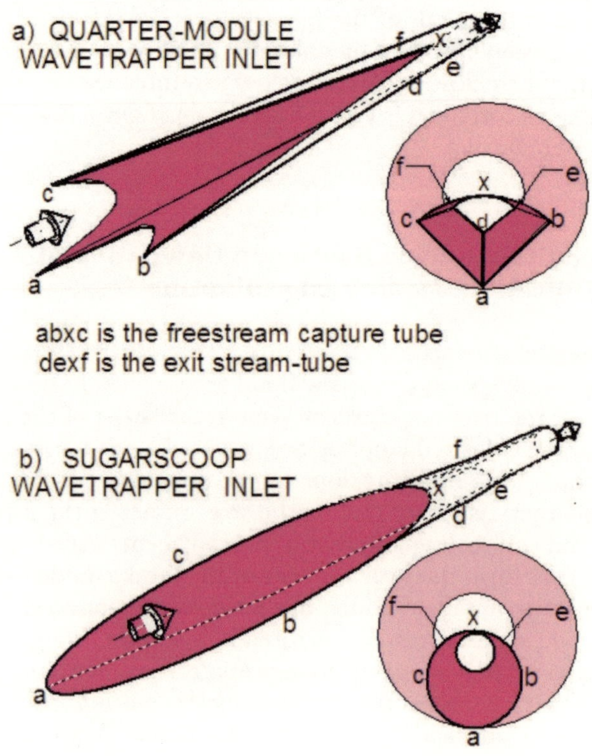

Figure 18.
Wavecatcher intake modules traced from full Busemann flows.

of the Busemann streamline. The scaling factor measures how far the streamline is from the axis of symmetry; its parameter ϕ is unique to each streamline, being the circumferential location of the streamline, the azimuthal angle, measured around the axis. It defines the cross-sectional shape $y(\phi)$ of the freestream capture tube. Note that, on the resulting surface, the variable θ uniquely determines all property values including surface inclination—this being a characteristic of conical symmetry.

Two streamline traced intake modules are shown in **Figure 18**. Both are based on Busemann flow. In **Figure 18a**, the freestream capture tube shape is a quarter circle. The exit is also a quarter circle. Four such modules were placed back-to-back to construct the intake in **Figure 19**. Such four-module intakes were tested in a gun tunnel at Mach 8.33 [4] and this intake, on a scramjet, was launched from a ballistic gun at Mach 5 [41]. **Figure 18b** shows an intake, also traced from Busemann flow,

Figure 19.
Four-module Mach 5 scramjet intake based on Busemann flow.

from a circular capture tube shape, where the exit shape is also circular. Such a module was tested in a wind tunnel at Mach 4 [25].

The swept leading edges of modular wavecatcher surfaces permit flow spillage at design and below-design conditions thus promoting intake flow starting. Once started, the apparent three-dimensional intake flowpath contains a started, steady flow with the original Busemann flow properties. These are the two significant virtues of wavecatcher intake modules.

7.2 Morphing (modification of intake flow cross section)

The technique of generating wavecatcher intakes, described in Section 7.1, produces exit flow cross-section shapes that are geometrically similar to the freestream capture streamtube shapes. The purpose of morphing is to produce cross-sectional shapes of the intake flow path that gradually transform the intake's entry shape to a geometrically different exit shape while, as much as possible, preserving the cross-sectional areas as well as the flow characteristics. For example, the flow from a quarter-circle entry is to be morphed to feed a circular combustor.

Figure 20 shows three orthogonal views of a wavecatcher intake and its cross sections when morphed from a quarter-circle to a full-circle. A detailed morphing method, as applied to the Busemann intake streamline $r = f(\theta)$, is pictured in **Figure 21**.

We illustrate by morphing a large, square (blue) inflow cross section into a (red), small circular exit section. A typical morphed Busemann intake design starts from specifying the initial conditions at the Busemann shock. A Busemann streamline $r = f(\theta)$, **Figure 3**, is then calculated from the shock to the freestream, as in Section 3. For each value of ϕ, ranging from 0 to 360°, in a meridional plane, two streamlines are calculated, $r_1 = y_1(\phi)f(\theta)$ and $r_3 = y_3(\phi)f(\theta)$ where $y_1(\phi)$ is the distance from the axis to the freestream capture cross section (blue) and $y_3(\phi)$ is the distance from the axis to the exit flow cross section (red). All the r_1 streamlines project downstream from the leading edge and all the r_3 streamlines project upstream from the trailing edge. The morphed streamline shape, $\bar{r} = \bar{r}(\phi, \theta)$, is then composed of the weighted average of the two streamlines, $\bar{r} = r_1 + g(\theta)[r_3 - r_1]$ where $g(\theta)$ is some assigned morphing function that varies from 0 to 1 as θ varies from the freestream Mach angle to the shock inclination. $\bar{r} = \bar{r}(\phi, \theta)$, then, represents a streamtube surface that joins the square leading edge to the circular trailing

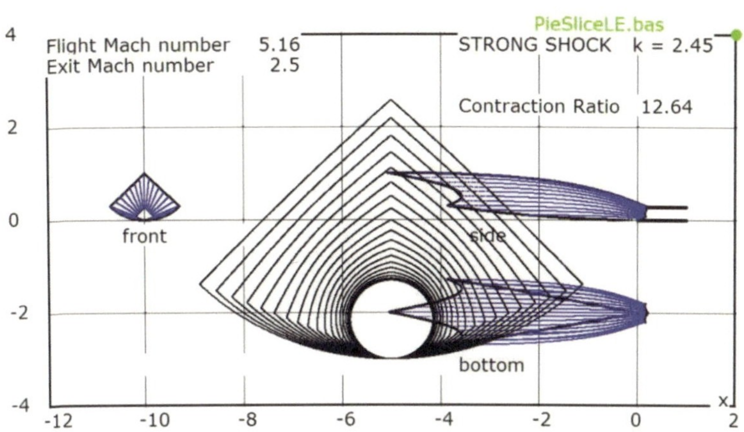

Figure 20.
Three (blue) orthogonal views of a wavecatcher module and cross sections of the modular intake (black) when morphed from a quarter circle to a full circle.

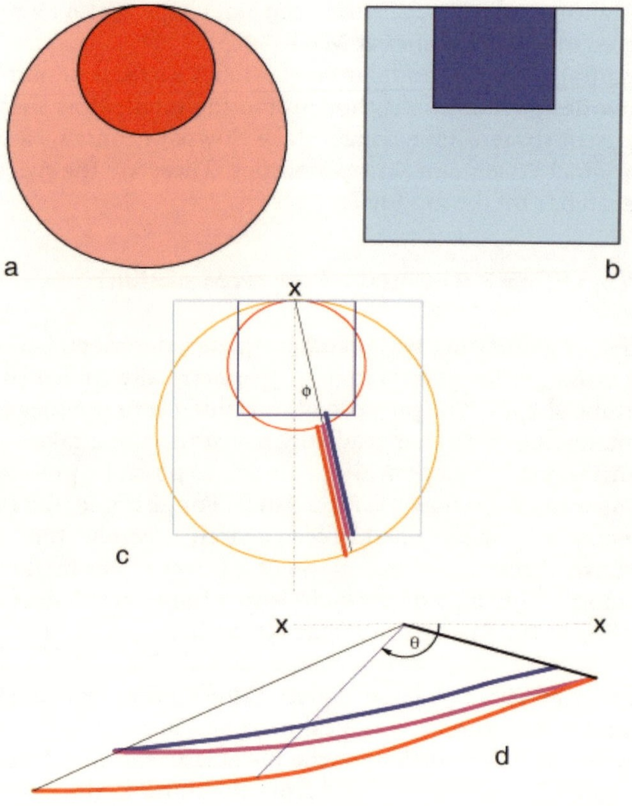

Figure 21.
Morphing of streamline-traced square (blue) and circular (red) streamlines into composite (purple) yielding cross section transition from large blue square to small red circle: (a) is exit geometry; (b) is entry geometry; (c) shows front view of streamlines; and (d) shows side view of streamlines.

edge, as shown in **Figure 21**. The surface grid points are easily calculable from $\bar{r} = \bar{r}(\phi, \theta)$ where $0 \leq \phi < 2\pi$ and $\theta_2 \leq \theta \leq \mu_1$ and the Cartesian coordinates of the surface are:

$$x = \bar{r}\cos\theta \qquad y = \bar{r}\sin\theta\cos\phi \qquad z = \bar{r}\sin\theta\sin\phi$$

Morphing can be used also if the axes of the entry and exit flows are offset, but still parallel.

Although morphing is applied to Busemann flow streamlines, Busemann flow is not preserved in the morphed intake. The morphing process is a purely geometric exercise and its arbitrary nature makes it necessary to verify the morphed intake's flow features and performance, by CFD or experiment. VanWie et al. [7] examined the results of applying various weighting functions and calculated the performance of the morphed intakes using CFD.

7.3 Intake foreshortening: truncation and stunting

As discussed in Section 1, full Busemann intakes are inherently long and hence subject to substantial viscous losses and high structural weight. An examination of the Busemann intake flow-field reveals that the surface at the leading edge has no deflection or curvature in the streamwise direction, presenting no compression of the ingested freestream flow. Thus the leading surface makes little contribution to the task of compressing the flow in the intake. Even worse, it supports a boundary layer with high shear and attendant losses of intake efficiency. There is then a good reason to expect an improvement in efficiency as a result of eliminating the leading edge surface by foreshortening the intake surface. At the same time one can expect a deterioration of efficiency because the foreshortened intake now has a positive deflection generating a leading edge shock that produces an efficiency loss in the inviscid flow. There is a design trade-off here, between boundary layer and shock losses, which arises from intake foreshortening and it becomes of interest to find an amount of intake foreshortening that minimizes the sum of the boundary layer and the shock losses—maximizes the efficiency. This section describes two representative geometric methods of achieving foreshortening of air intakes, *truncation* and *stunting*. *Truncation* shortens the intake by removing some part of the leading edge surface. The effect of truncation of the Busemann intake was studied in [10, 16, 18]. *Stunting* is longitudinal contraction of the Busemann intake achieved by multiplying all streamwise intake surface coordinates by a constant factor <1. This is linear stunting or telescoping. When applied to a Busemann intake profile, the intake is foreshortened while the flow areas and the zero leading edge flow deflection and curvature are retained. No shock is produced at the leading edge and the overall design contraction is not changed.

CFD-generated intake performance data is presented for a Mach 8, full Busemann intake, flying at an altitude of 30 km, when foreshortened by various amounts of truncation or stunting. **Figure 22** is a plot of intake total pressure recovery against fractional foreshortening of the full Busemann intake calculated by a Navier-Stokes code. The Busemann intake, with applied boundary layer and terminal shock losses lead to a total pressure recovery of 42% for the un-shortened intake.

The effect of truncation on total pressure recovery by various amounts of truncation is shown by the blue curve in **Figure 22**. Truncation produces a modest increase of total pressure recovery from 42 to 46% at near 30% truncation and it appears that intake efficiency is not very sensitive to the amount of truncation.

The effect of stunting, on total pressure recovery, by various amounts is shown by the red curve in **Figure 22**. Total pressure recovery peaks at 47% near 15% foreshortening; decreasing noticeably as stunting increases.

Assessment of truncation and stunting. Both truncation and stunting produce only modest, 4 and 5%, improvements in intake efficiency. However, since the methods are geometrically different, they affect intake capability differently as shown by the compression and contraction ratios in **Figure 23**.

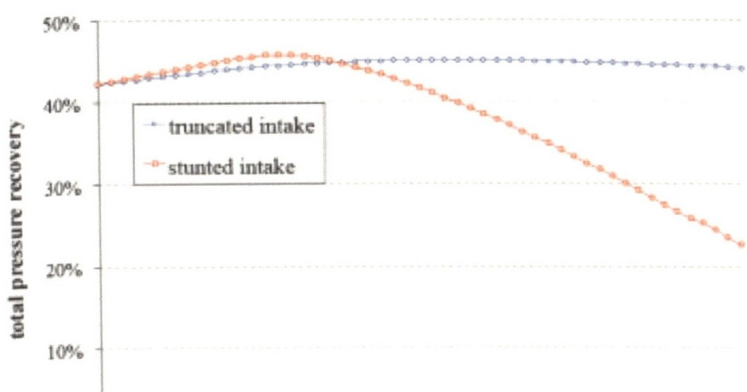

Figure 22.
Effects of truncation and stunting on Busemann intake total pressure recovery.

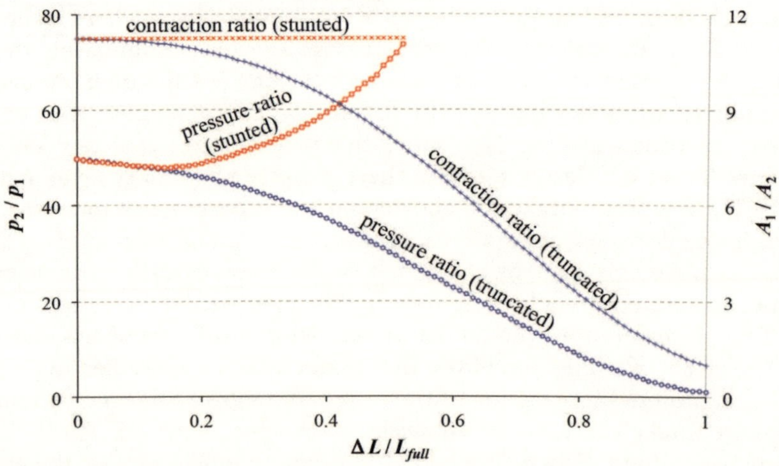

Figure 23.
Variation of intake compression and contraction as caused by truncation and stunting.

Intake performance, both efficiency and capability, are not affected much by considerable amounts of either truncation or stunting. This is due to the fact that the high-loss leading edge boundary layer flow is not eliminated but merely moved downstream. Also there is some increase in inviscid flow losses from the finite angle leading edge from truncation. However, an estimated 15–30% weight saving is available through wall materials elimination resulting from intake fore-shortening. It appears that the significant advantage of truncation and stunting is not to intake performance but to the saving of structural weight. Similar results were found in [10].

7.4 Leading edge blunting

Busemann flow has no deflection at the leading edge so that the leading edge tends to be sharp and thin. Such leading edges are difficult to cool at hypersonic speeds. Transpiration cooling is made possible by a slight rounding of the leading edge. Rounding or blunting affects both the viscous as well as the inviscid flow in the intake [10]. The strong bow shock causes a hot entropy layer to overlay the boundary layer and cause it to thicken. The same shock focuses on the symmetry axis producing a Mach reflection at the center line. It was shown in [10] that a 1 mm diameter leading edge on a 500 mm diameter Busemann intake, flying at Mach 10 and 30 km altitude, is optimal in reducing the viscous and inviscid losses. It seems that the combination of blunting and stunting should be such that the conical shock is kept incident on the Busemann surface corner, so that no reflected shock waves are formed, keeping the exit flow uniform.

8. Startability

The Kantrowitz criterion for intake starting [26] says that the normal shock, in front of an intake duct, will move downstream and out of the duct if the flow at the exit of the duct is not choked—the duct flow will start. This criterion applies to the normal shock at the entry of the duct as well as at any other position in the duct. On a wavecatcher intake, **Figures 18** and **19**, overboard flow spillage will allow the shock to move downstream, over the external/open portion of the intake, until it reaches the V-notch at the beginning of the internal flow. This is made possible by an effectively large flow area, on the moving, post-shock side, allowing overboard flow spillage. If the Kantrowitz condition, for the shock at the point of inflection, at the V-notch, is satisfied the shock will continue moving downstream, out through the internal flow section, and the intake will start.

The entry area of the internal flow, A_f, is defined as the conical surface at the angular position where the surface is inflected because the flow is normal to the conical surface at the inflection and a stationary, conical, normal shock is compatible with the flow there. The size of this area is available from a Busemann intake calculation. This area is needed for application of the Kantrowitz starting criterion.

8.1 Startability of the weak shock Busemann

Note that, for given M_2 and δ_{23}, Eq. (9), gives two solutions for θ_{23}, for a weak and a strong shock. This leads to the possibility of generating two different Busemann intakes, the weak shock version would have supersonic and the strong shock would have a subsonic exit flow. Because of its supersonic exit flow the weak shock intake is better suited for scramjet application. However, at contractions to be useful for scramjets, the Busemann intake with a weak shock does not start spontaneously or if it does start then it does so for intakes with an insufficient amount of contraction.

The determination of startability for a wavecatcher Busemann intake is as follows. At first we examine the startable weak-shock Busemann flow to show that it does not provide sufficient compression:

a. At a prescribed **weak** shock angle, pre-shock Mach number and exit radius y_3 begin integrating the T-M Eqs. (5) and (6) towards the upstream;

b. Halt the integration when reaching the inflection point, (of) in **Figure 3**, where $u = 0$. Note the Mach number $M_f = v$ and $\overline{\theta} = \theta$ at this point (u is zero here);

c. Calculate surface area of the conical surface at inflection point, A_f;

d. Apply the Kantrowitz criterion to M_f, A_f and A_3 to determine if the internal passage will start.

e. If a start is indicated the intake is practical and integration can be continued to find the freestream (entry) Mach number, M_1, and the other overall performance parameters such as the exit-to-entry area ratio A_3/A_1, the compression ratio p_3/p_1 and the total pressure recovery p_{t3}/p_{t1}.

Many such calculations, starting from *weak shock waves*, (os) in **Figure 3**, were performed with the outcomes plotted on a graph of area ratio, A_3/A_1 *vs.* entry Mach number, M_1, in **Figure 24**. Each result is shown as a dot that is colored green if the totally internal flow Busemann intake duct starts, green or yellow if the wavecatcher Busemann intake module starts (as determined in d) above) and red if there is no start. Curves of the "startability index," $S = (A_1 - A_i)/(A_K - A_i)$, measure the location of a dot on the overall area ratio scale where $S = 0$ on the isentrope and $S = 1$ on the Kantrowitz criterion. Intermediate, fractional values, are on curves between these limits. The curve for $S = 0.6$ seems to well represent the startability limit for wavecatcher Busemann intake designs based on the weak shock condition. As seen from the figure the wavecatcher design lowers the startable area ratio from about 0.6 to 0.4. This is still not good enough. For good engine performance, it is

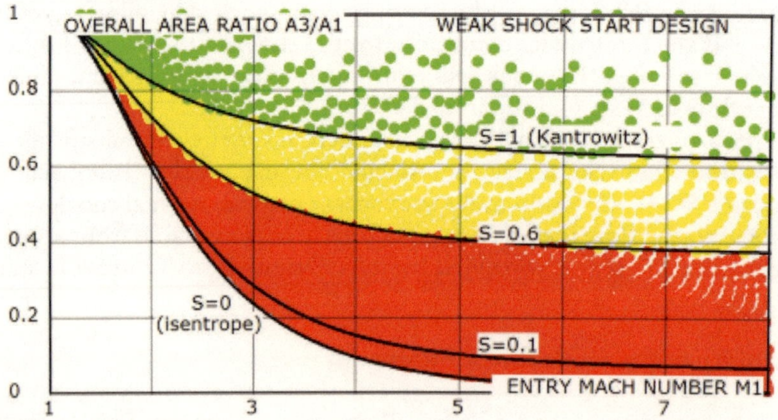

Figure 24.
Busemann intake startability based on weak shock design: green—full Busemann start; green and yellow—wavecatcher start with spillage; and red—no start.

desirable to operate intakes near the curve S = 0.1 in **Figure 24**. Startable weak shock Busemann designs do not produce this amount of contraction—the prospect of obtaining startable full or wavecatcher intakes, with sufficient compression, from weak shock Busemann flows is disappointing.

8.2 Startability of the strong shock Busemann

Startability calculations for the full and wavecatcher Busemann intakes were done also by starting the integration of Eqs. (5) and (6) from strong shock waves, following the (a–e) steps above.

The outcomes are plotted in **Figure 25**. Each result is shown as a dot that is coloured green if the totally internal flow Busemann intake duct starts, green or yellow if the wavecatcher Busemann intake module starts (as determined in d) above) and red if there is no start.

The strong shock version has high overall contraction but low internal contraction so that it will self-start at overall contractions useful to the scramjet as a wavecatcher, but the strong shock and subsonic exit flow are not useful to the scramjet engine. In a wavecatcher module the about-to-start strong shock will be held in place by the appropriate back-pressure. It will move downstream if the

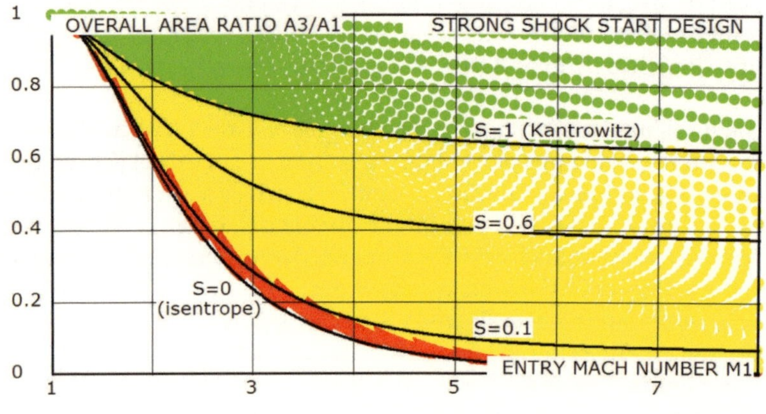

Figure 25.
Busemann intake startability regimes based on strong shock design. green—full Busemann start; green and yellow—module start with spillage; and red—no start.

back-pressure is lowered to be replaced by a weak shock structure. This now opens a possibility. We calculate a Busemann intake module, with a strong shock, such that it starts spontaneously, at a high overall contraction ratio, and then reduce the back-pressure to remove the strong shock and obtain a supersonic exit flow, with a weak shock. This yields an exit flow which is suitable for scramjet operation. In doing this, we note that the flow, from the freestream to the location of the strong shock has not changed as we switch to the weak shock, so that the intake remains on the strong-shock design flow up to the corner while being started. Also, the amount of internal contraction remains the same and we could really start the intake with the weak shock structure in the first place. The strong Busemann shape is really a design tool which leads to a modified Busemann flow but with a started intake of high compression and efficiency having a supersonic exit Mach number-an intake with a high overall contraction but with a low **internal**, startable contraction. The supersonic exit flow is no longer conical although its axial symmetry is preserved.

In comparison to the weak shock case (**Figure 24**) there is a considerable enhancement of startability in the enlarged yellow domain so that starting can be expected near the S = 0.1 curve, which is acceptable for scramjet applications.

The reason for this improvement in startability stems from the fact that, for the strong shock option, the angular distance between the strong shock (at the corner) and the inflection cone is small so that A_f and A_3 areas are close in size, i.e., there is not much internal contraction. This makes it easy for the conical normal shock, at the inflection location, to be swallowed. The strong shock design's aim is to produce a wavecatcher Busemann intake with a high overall contraction ratio and a low internal contraction that starts spontaneously. The calculated shape is compatible with a normal conical shock positioned at the inflection location (angle) and we select the intakes, with internal contractions, that permit the shock to be swallowed. Flow downstream of the inflection shock is subsonic. The strong shock is present only fleetingly during flow starting. After swallowing a weak shock system appears in the exit with supersonic flow downstream. This flow is suitable for a scramjet combustor. Within the calculated strong shock contour the supersonic weak shock flow is no longer conically symmetric and has to be examined via CFD and experiment. The end result is a wavecatcher Busemann intake with a high overall contraction (S $\sim$ 0.1) but with a low internal, self-starting contraction and hence a startable intake and supersonic exit flow.

Figure 26 is a schlieren picture of four quarter-circle Busemann intake modules each with flows started by the unsteady gun tunnel starting flow [4]. Arrow points

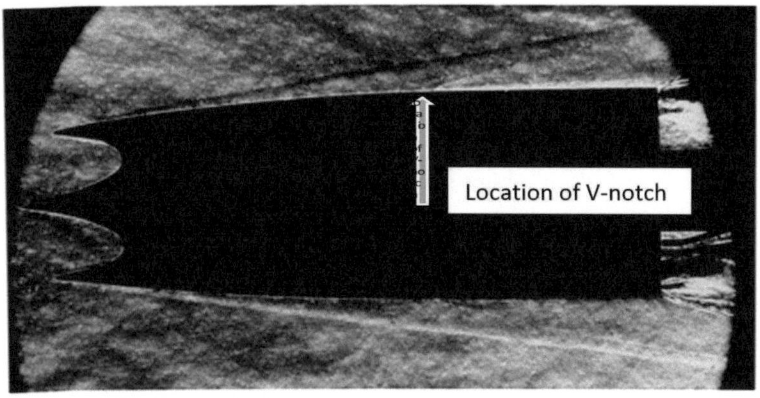

Figure 26.
*Impulsively started 4-module Busemann intake in the gun tunnel at Mach 8.33 (model is similar to **Figure 19**).*

to weak shock emanating from the v-notch in the cowl of one of the modules. Exit Mach number is 5.23.

9. Concluding remarks

A review and summary is presented of hypersonic air intake technology highlighting design objectives, basic flows, airframe integration, flowpath modification and intake flow startability. Taylor-Maccoll equations and Busemann flow are presented as the basis for constructing modular Busemann intakes. The first-order Taylor-Maccoll equations are recast with Mach number components as dependent variables. These equations illustrate the free-standing conical shock, the axisymmetric centered compression fan, characteristics, surface curvature and inflection point, surface pressure gradient, surface Mach number gradient and gradients at conical shock waves. A chart is presented for assessing the performance of the Busemann-flow-based intake on the basis of its capability to reduce Mach number and its efficiency as measured by total pressure recovery. Experimental results indicate that viscous effects cause the total pressure recovery to drop from near 1 to 0.5 for a Busemann intake at Mach 8 and 30 km flight altitude.

Wavecatching (streamline tracing), morphing and foreshortening, as attempts at conditioning and improving the performance of the basic Busemann intake flow, are presented to show that (a) wavecatching is a useful technique to create modular startable intakes; (b) morphing is useful in integrating the intake shape with other geometric requirements of the airframe and combustor; (c) foreshortening leads to minor gains in intake performance but large weight savings; (d) small amounts of leading edge blunting cause large changes in the intake's shock structure.

A novel, *strong shock* method is presented, that uses strong-shock boundary conditions for designing spontaneously startable, modular Busemann intakes of high performance. This analytical approach allows pre-determination of Busemann intake startability; offering great simplicity in the search for flowpath surface shapes that yield startable intakes with high compression, high efficiency and supersonic exit flows. This improved startability is made possible by the wavecatcher's ability to spill mass flow during external compression combined with the moderate contraction of the internal flowpath.

Busemann flow contains unique fluid mechanical features: (a) a flow passage from a uniform, high Mach number flow, to another uniform, lower Mach number flow; (b) internal, convergent flow with an inflected surface; (c) conical flow where high gradients are near the center line and milder gradients are at the walls; (d) an axisymmetric and conically symmetric centered compression fan; (e) a free-standing conical shock, bounding irrotational flow; (f) the last two features combining to preserve conical flow throughout. These are unique and fortuitous virtues, being significant in making the Busemann streamtube and its flow characteristics useful as a basis for designing high performance air intakes for hypersonic airbreathing engines.

Acknowledgements

Most CFD calculations were done by E.V. Timofeev, R.B. Tahir and Hideaki Ogawa. G. Emanuel provided useful comments. Many thanks.

Author details

Sannu Mölder
Ryerson University, Toronto, ON, Canada

*Address all correspondence to: smolder@ryerson.ca

References

[1] Busemann A. Die achsensymmetrische kegelige überschallströmung. Luftfahrtforschung. 1944;**19**(4): 137-144

[2] Courant R, Friedrichs K. Supersonic Flow and Shock Waves. New York: Interscience; 1948

[3] Molder S, Szpiro EJ. Busemann inlet for hypersonic speeds. Journal of Spacecraft and Rockets. 1966;**3**(8): 1303-1304

[4] Molder S, Romeskie JM. Modular hypersonic inlets with conical flow. In: AGARD Conference on Hypersonic Boundary Layers and Flow-fields; Proceedings, No. 30; Cranfield: NATO/ AGARD; 1968

[5] Van Wie DM. Scramjet inlets. In: Scramjet Propulsion. Washington, USA: American Institute of Aeronautics and Astronautics; 2001. pp. 447-511

[6] Grodzovskii GL. Supersonic axisymmetric conical flow bordering on a symmetrical parallel flow through a shock wave. RAE translation of Prikladnaya Matematika i Mekhanika. 1959;**XXIII**(2):379-383

[7] Taylor TM, Van Wie D. Performance analysis of hypersonic shape-changing inlets derived from morphing streamline traced flowpaths. In: 15th AIAA International Space Planes and Hypersonic Systems and Technologies Conference; 28 April 2008–1 May 2008; Dayton, OH: United States; 2008

[8] Duganov VV, Polyakov VV. Hypersonic flow in a converging conical channel. Izvestiya VUZ. Aviatsionnaya Tekhnika. 1976;**19**(2)

[9] Stockbridge RD. Design Requirements for Self-Starting Busemann Diffusers, BBP-78-1 Memo. Applied Physics Laboratory, Johns Hopkins University; 1978

[10] Drayna TW, Nompelis I, Candler GV. Hypersonic inward turning inlets: Design and optimization. In: 44th AIAA Aerospace Sciences Meeting 2006; 9–12 January 2006; Reno, NV: United States; 2006

[11] Molder S, Sullivan PA, McGregor RJ, Sislian JP, Paisley TW. Investigations in the fluid dynamics of scramjet inlets. USAF Contract Report F33615–87-C-2748 and JHU/APL Contract APL602235–0; 1992

[12] Van Wie DM, Molder S. Applications of Busemann inlet designs for flight at hypersonic speeds. In: Aerospace Design Conference; 3–6 February 1992; Irvine, CA, United States; 1992

[13] Matthews AJ, Jones TV. Design and test of a modular waverider hypersonic intake. Journal of Propulsion and Power. 2006;**22**(4):913-920

[14] Sheng YQ. Hypersonic inlet flow computation [MA Sc thesis]. UTIAS, University of Toronto; 1992

[15] O'Brien TF, Colviflet JR. Analytical computation of leading-edge truncation effects on inviscid Busemann-inlet performance. Journal of Propulsion and Power. 2008;**24**(4):655-661

[16] Ogawa H, Shoesmith B, Mölder S, Timofeev E. Viscous correction and shock reflection in stunted busemann intakes, shock wave interactions. Selected articles from the 22nd International Shock Interaction Symposium, University of Glasgow. United Kingdom; 2018

[17] Galkin VM, Zvegintsev VI, Vnuchkov DA. Investigation of annular supersonic inlets with isentropic

compression. Thermophysics and Aeromechanics. 2016;**23**(5):645-655

[18] Ogawa H, Mölder S, Boyce R. Effects of leading-edge truncation and stunting on drag and efficiency of Busemann intakes for axisymmetric scramjet engines. Journal of Fluid Science and Technology. 2013;**8**(2):186-199

[19] Ogawa H, Mölder S. Numerical analysis of hysteresis in mode transition of centerline mach reflection in stunted Busemann intakes for axisymmetric scramjet engines. JAXA Special Publication. 2014;**JAXA-SP-13-011**: 145-150 (in Japanese)

[20] Seddon J, Spence A. The use of known flow fields as an approach to the design of high-speed aircraft. In: Hypersonic Boundary Layers and Flow Fields; AGARD Conference Proceedings No. 30. London, UK: AGARD/NATO; 1968

[21] Keirsey JL, Snow ML. Modular inlet investigation. Quarterly Report AQR/66–1, Aeronautics Division, R&D, APL, Johns Hopkins University; 1966

[22] Smart MK. Calculation of stream-traced hypersonic inlet performance on and off design. In: 21st International Symposium on Shock Waves; Great Kepple Island, Australia; 20-25 July. 1997

[23] Smart MK. Design of three-dimensional hypersonic inlets with rectangular-to-elliptical shape transition. Journal of Propulsion and Power. 1999;**15**(3):408-416

[24] Stockbridge RD. Parametric study of busemann inlets, BBP Memo 77-128, APL, Johns Hopkins University; 1977

[25] Jacobsen LS, Tam C-J, Behdadnia R, Billig FS. Starting and operation of a streamline-traced Busemann inlet at Mach 4. In: AIAA/ASME/SAE/ASEE 42nd Joint Propulsion Conference; 9–12 July 2006; Sacramento, CA, United States; 2006

[26] Kantrowitz A, Donaldson CD. Preliminary Investigation of Supersonic Diffusers. NACA-ACR-L5D20. 1945

[27] Molder S, Timofeev E, Tahir R. Flow starting in high compression hypersonic air inlets by mass spillage. In: 40th AIAA/ASME/SAE/ASEE Joint Propulsion Conference and Exhibit; Fort Lauderdale FL; AIAA 2004-4130. 2004

[28] Ames Research Staff. Equations, Tables and Charts for Compressible Flow. NACA Rep. 1135; 1953

[29] Kreyszig E. Differential Geometry. Mineola, USA: Dover; 1959

[30] Dalitz RH. Some mathematical aspects of compressible flow, Report ACA-20, Australian Council for Aeronautics; 1946

[31] Henderson LF. A critique of the starting phenomena on supersonic intakes. Zeitschrift für Flugwissenschaften.15, Heft. 1967;**15**:2

[32] Najafiyazdi A, Tahir R, Timofeev EV, Molder S. Analytical and numerical study of flow starting in supersonic inlets by mass spillage. In: 43rd AIAA/ASME/SAE/ASEE Joint Propulsion Conference; 8–11 July 2007; Cincinnati, OH, United States; 2007

[33] Timofeev EV, Tahir RB, Molder S. On recent developments related to flow starting in hypersonic air intakes. In: 15th AIAA International Space Planes and Hypersonic Systems and Technologies Conference; Dayton OH; AIAA 2008-2512

[34] Tahir R. Starting and unstarting of supersonic intakes [M.Eng. thesis]. Toronto, Canada: Ryerson University; 2003

[35] Tahir R. Analysis of shock dynamics in supersonic intakes [thesis]. Montreal, Canada: McGill University; 2008

[36] Studzienny A, Molder S, Timofeev E, Voinovich P. Starting of a perforated supersonic inlet: A CFD simulation. In: Proceedings of the 22nd International Symposium on Shock Waves. Imperial College, London UK; 18-23 July. 1999

[37] Tahir RB, Molder S, Timofeev EV. Unsteady starting of high mach number air inlets—A CFD study. In: 39th AIAA/ASME/SAE/ASEE Joint Propulsion Conference and Exhibit; 20–23 July 2003; Huntsville, AL, United States; 2003

[38] Molder S. A benchmark for internal flow CFD code. Computational Fluid Dynamics Journal. 2003;**12**(2):408-414

[39] Walsh P, Tahir R, Molder S. Boundary-layer correction for the Busemann hypersonic air inlet. Canadian Aeronautics and Space Journal. 2003;**49**(1):11-17

[40] Mölder S, Timofeev EV, Lesage EV, Pimentel R. Free standing conical shock. In: Proceedings of 28th International Symposium on Shock Waves, Manchester, England; 2011

[41] Bernhart R. Supersonic martlet/scramjet set for first flight in Canada. Technology World. May 8, 1967

[42] Emanuel G. Analytical Fluid Dynamics, (3rd edition). Boca Raton FL, USA: CRC Press; 2000

[43] Anderson JD. Modern Compressible Flow: With Historical Perspective. New York: McGraw-Hill; 1990

[44] Thompson PA. Compressible Fluid Dynamics. New York: McGraw-Hill; 1971

[45] Ralston A, Wilf HS. Mathematical Methods for Digital Computers. Toronto: John Wiley and Sons; 1960

[46] Zucrow MJ, Hoffman JD. Gas Dynamics. New York: Wiley; 1976

[47] Shapiro AH. The Dynamics and Thermodynamics of Compressible Fluid Flow (Vol 2). New York: Ronald Press; 1954

[48] Mölder S. Curved shock theory. Shock Waves. 2016;**26**(4):337-353

[49] Flock AK, Guelhan A. Viscous effects and truncation effects in axisymmetric Busemann scramjet intakes. AIAA Journal. 2016;**54**(6):1881-1891

[50] Greene F, Hamilton H. Development of a boundary layer properties interpolation tool in support of orbiter return to flight. In: 9th AIAA/ASME Joint Thermophysics and Heat Transfer Conference; 2006

[51] Mcnally WD. BLAYER—Compressible laminar and turbulent boundary layers in arbitrary pressure gradients, Computer Program, LEW-11097. NASA Lewis Research Center; 1994

[52] Carter JE. A new boundary-layer interaction technique for separated flows. NASA-TM-78690. 1978

Section 3

Flowfield Investigations

Chapter 6

Numerical Simulation of Base Pressure and Drag of Space Reentry Capsules at High Speed

Rakhab C. Mehta

Abstract

The numerical simulations over several reentry vehicles are carried out by solving time-dependent compressible laminar axisymmetric Navier-Stokes equations for Mach 1.2–6.0. The fluid dynamics equations are discretized in spatial coordinates using integral formulation in conjunction with a finite volume method which reduce to semi-discretized ordinary differential equations. A local time-step is used to achieve steady-state solution. The numerical computation is carried out on a single-block structured computational grid. The flowfield features over the reentry vehicle such as formation of a bow shock wave ahead of the fore-body, expansion fan on the shoulder, and recirculation zone in the base region are well captured in the numerical simulations. Lower pressure acting on the base of the reentry capsule acts as base drag. The base drag coefficient based on maximum cross-section of the reentry capsule must satisfy inequality. The base drag coefficient is a function of several geometrical parameters of the fore-body and back-shell of reentry capsule, boundary layer, formation of free-shear layer in the wake region and freestream Mach number. The purpose of this chapter is to numerically evaluate and tabulate the base pressure and the base drag coefficients of various reentry space capsules at zero angle of incidence.

Keywords: aerodynamic, base drag, CFD, high speed flow, viscous flow, reentry vehicle, shock wave

1. Introduction

A space vehicle may be designed with several trajectory options such as non-lifting (steep or shallow), lifting (skipping or diving), terminal (gravity assist), thrusting (jet-on) reentry. The base pressure and heat flux are of paramount importance for smooth deployment of parachute and successful landing of a spacecraft. Cassanto [1] has carried out a number of wind tunnel and free-flight experiments to obtain the base pressure. Lamb et al. [2] have reviewed the base pressure on the reentry vehicle at high speed, which depends on wake flow characteristic, freestream conditions and edge properties of boundary layer at the shoulder of the module. The base pressure correlation for supersonic flows are compared by Kawecki [3] using the ground test data and with different vehicles such as ABC, MK-3, 4, 12, MTV, reentry F, REX, RVTO, SAMAST, TVX and WAC. A supersonic analysis of the SPR INT blunted cone-flare is carried out by Terry and Barber [4] employing computational fluid dynamics (CFD) method as well as wind-tunnel

IntechOpen

testing at Mach 3. Experimental and numerical computations by Togiti et al. [5] of the flow behind a truncated cylinder in a supersonic flow reveals almost constant base pressure coefficient. Base flow investigation of the Apollo AS-202 is presented in detail by Walpot et al. [6].

The bow shock wave is formed ahead the blunt body which is enclosed by a subsonic-supersonic region between the blunt body and the bow shock wave. The wall pressure distribution, the location of the sonic line and shock stand-off distance on the spherical cap region have been analytically analyzed by Chester [7] and Freeman [8] at very high speeds with an adiabatic index near to unity which predicts a singular point at 60° from the stagnation point. However, the analytical approach [9] for the high-speed flow over the blunt-body is found to be the most difficult and complex. The flowfield over the reentry capsule becomes further complicated due to the presence of bevel at the shoulder and shape of the base shell of the reentry module.

Aerodynamic analyses of the COMmercial Experiment Transport (COMET) reentry capsule have been carried out by Wood et al. [10] solving the thin layer laminar Navier-Stokes at high speeds. Yamamoto and Yoshioka [11] have performed flowfield computation over the Orbital Reentry EXperiments (OREX) using CFD method in conjunction with flight aerodynamic data. Ivanov [12] cataloged different shapes for non-winged reentry vehicles. The aerodynamic characterization of the CARINA reentry module in the low supersonic Mach regimes has been performed employing numerical and experimental methods [13]. The flowfield simulations over the Beagle-2 spacecraft have been obtained by Liever et al. [14] using CFD code for low supersonic to hypersonic speeds. Mehta [15] has numerically simulated flowfield over atmospheric reentry demonstrator (ARD) and space recovery experiment (SRE).

Wind tunnel testing of the orion crew module (OCM) has been carried out by Ross et al. [16] to obtain the aerodynamic forces. Murphy et al. [17] have presented experimental static aerodynamic data for the OCM reentry capsule and analyzed with the help of surface flow visualization and computational results. Shape optimization design method has been presented by Zhenmiz et al. [18] for the conceptual design of reentry capsules. Ali et al. [19] have studied effects of nose-bluntness ratio on the aerodynamic performance of reentry capsules. CFD analyses of space vehicle are performed employing H3NS and FLUENT code by Viviani et al. [20] to analyze the flowfield over various capsules. Chen et al. [21] have carried out numerical simulations of flowfield for aerodynamic design of reentry capsules. Weiland [22] has presented aerodynamic characteristics of several non-winged capsules. Effects of geometrical parameters over fore-body of various reentry vehicles have been numerically investigated at high speeds [23, 24].

The flowfield feature of shock wave interaction over a double-cone module includes a local flow separation attributed to the semi-cone angle of the double-cone configuration. It has been also observed that these flowfields are controlled by the vorticity in the incoming boundary layer and the strength and the orientation of the shock wave. Numerical and experimental studies have been performed by many researchers [25–27].

It is worth to mention here that considerable difficulties encountered for obtaining aerodynamic data from wind-tunnel testing are attributed to model-sting interference effects. The shock tunnel is having short duration of testing time. In free flight experiments, a scaled model is launched inside a range and orthogonal shadowgraphs are taken as the capsule flies by each shadow graph station. The CFD approach provides flowfield behaviour and aerodynamic coefficients without the sting interference effect. In the present Chapter, numerical studies were undertaken for a freestream Mach number range of 1.2–6.0. The numerical simulation is to solve the axisymmetric laminar compressible Navier-Stokes equations on a single-block structured grid, i.e., the number of grid points in the radial direction in each zone of the computational region is same. Surface pressure variations over the vehicles are computed which reveal a systematic understanding of the flow features over the capsule at high speeds. It also reveals the effect of geometrical parameters on aerodynamic base drag coefficient. The unsteady flow characteristics of the OREX and the Beagle-2 are analyzed in Ref. [28].

2. Review of the base pressure data

A base pressure experiment for determining the atmospheric pressure profile of planets applicable for Mars, Venus and Jupiter entry probe is presented by Cassanto [29]. The results of a reentry vehicle flight test have demonstrated by Nieden et al. [30] for feasibility of the experiment to obtain the atmospheric pressure profile. The fore-body shape of the reentry vehicle affects the base pressure [31]. Cassanto [32] wind-tunnel data with a sting attached to a model at Mach 4 predict 25–50% higher than the flight data. The Euler code (SAN DIAC) has been employed to compute flowfield over a large number of space vehicles by Noack et al. [33]. Comparisons were made between numerical and experimental results by McWherter et al. [34] using parabolized Navier-Stokes code SPRINT.

Theoretical studies of the fluid dynamics in the base flow region of the vehicle were presented by Baum [35]. They found in the analysis that the outside flow $M > 1$ is distinguished from relative low velocity core $M < 1$ of the base flow regimes by a separated flow.

The base pressure for sphere-cone configuration [36] at zero angle of attack was found to be a strong function of cone-angle and bluntness ratio. Analysis of flight-test base pressure data [37] for 10° sharp-cone has shown radial base pressure gradient in laminar flow. It is experimentally found that the base pressure is function of Reynolds number under laminar flow condition. Cassanto et al. [38] have investigated local flow effects on base pressure for the 10° sharp-cone configurations. Free-flight base pressure obtained using telemetry technique was compared with the sting-supported wind-tunnel data at Mach 4. Effects of Mach number on ratio of the base to freestream pressure (p_B/p_∞) in laminar and turbulent case are carried out in wind-tunnel and free-flight testing [38]. Correlation of free-flight base pressure data for Mach 4–19 has been obtained by Cassanto et al. [39]. After-body configuration [40] affects the base pressure ratio levels by about 25% compared to experimental studies. Base pressure measurements on slender cones at zero angle of attack with laminar flow condition on after-body were presented for Mach 11.9. Full-scale flight test base pressure results for a blunt planetary entry probe configuration having a blunt body 52° sphere-cone are analyzed by Cassanto [29]. The base pressure experiment is applicable for Mars, Venus, Jupiter reentry probe missions. The base pressure measurements on a 9° semi-cone angle at Mach range of 3.50–9.20 have been carried out by Zarin [41]. Flight-test base pressure measurements were conducted by Bulmer [42] for Mach number range of 0.5–15. The shapes of the Viking, Mars Path Finder (MPF), Mars Exploration Rovers (MER), Phoenix, Mars Science Laboratory (MSL), Mu-Science Engineering Satellite (MUSES-C) are similar to the Apollo capsule [43]. Aeroassist Flight Experiment (AFE) configurations have been analyzed numerically using two different Navier-Stokes flow solvers by Venkatapathy et al. [44]. The effect of base flow at low supersonic speeds on the sonic line location at hypersonic speed on aerodynamic coefficients has been analyzed by Gnoffo et al. [45]. Tam [46] and Menne [47] have computed flowfield over Viking, Bioconic and AFE vehicles employing Euler flow solver. A spherical blunt-cone/flare delft aerospace recovery test (DART) configuration is numerically analyzed by Otten [48] solving a laminar Navier-Stokes equations. Numerical analysis over blunted cone flare has been carried out at Mach 6 by Savino et al. [49]. Barnhardt [50] has carried out numerical simulation of flowfield in the wake region of a reentry vehicle at high speeds. The EXPeriment and Recovery of Space System (EXPRESS) reentry capsule at transonic and supersonic speeds is studied experimentally by Suzuki, and Abe [51].

It is important to state here that the base pressure can never be less than zero. The base pressure coefficient can be expressed as

$$C_{PB} = \frac{-p_\infty}{\frac{1}{2}\rho_\infty V_\infty^2}\left(1 - \frac{p_B}{p_\infty}\right) \tag{1}$$

Lower pressure is acting on the base experiences another form of aerodynamic base drag. The base drag coefficient based on the maximum cross-section of the reentry space capsule must satisfy inequality

$$C_{DB} < \frac{2}{\gamma M_\infty^2} \tag{2}$$

Thus, it can be noticed that the base pressure is having complex flow features which are a function of several variables such as geometrical parameters of the fore- and after-body of the reentry space vehicle, Mach number and Reynolds number. The measurements of base pressure in the wind-tunnel testing are affected by the presence of the sting attachment to the model. The free-flight experiment needs pneumatic launcher mechanism, pressure transducer, motion picture photography equipment, antenna, receiver and recording devices. However, the base pressure data obtained from the free-flight experiments are not affected by the sting attachment to the model as in the wind-tunnel testing. The numerical simulations are most suitable and inexpensive tool to evaluate flow characteristics, base pressure and drag coefficient for wide range of Mach numbers and Reynolds numbers.

3. Glimpse of flowfield over reentry vehicles

The flowfield features over the reentry capsule can be delineated through the experimental and theoretical investigations at high speed. The nomenclature and geometrical parameters of a typical reentry capsule is depicted in **Figure 1(a)** that leads to the necessity to investigate the influence of the geometrical parameters such as D, α_N, R_C, α_B, and L on the flowfield and aerodynamic characteristics. A schematic sketch of flowfield is delineated in **Figure 1(b)** based on shadowgraph and schlieren pictures. The significant flow features are described by the following. In the fore-body section of the capsule, the fluid decelerates through the bow shock wave depending on the cruise speed and altitude. At the shoulder of the capsule, the flow turns and expands rapidly and boundary layer detached, forming a free-shear layer in the back-shell region that separates the inner recirculating flow region behind the module from the outer inviscid flowfield. The latter is recompressed and turned back to the freestream direction, first by the so-called lip shock wave, and further downstream by the recompression shock wave. At the end of the recirculating flow past the neck, the free-shear layer develops in the wake trail. A complex flow structure often includes a lip shock wave associated with the beveled expansion fan and wake trail adjacent to the shear layer confluence. The corner expansion process is an expansion fan pattern changed by the presence of the approaching boundary layer and radius of the bevel or shoulder, R_C. The wake flow features

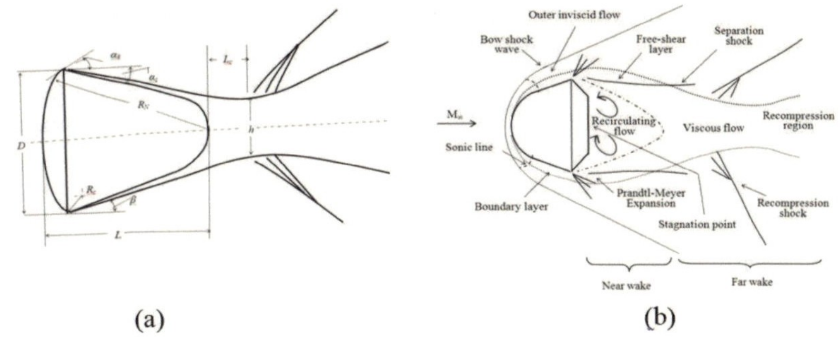

Figure 1.
Representation (a) geometrical parameters (b) flow features.

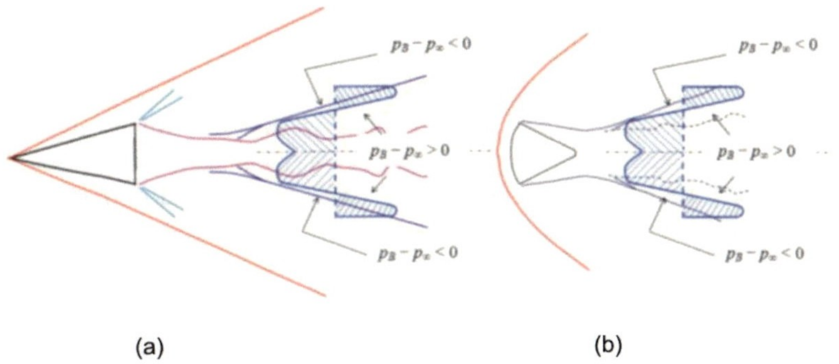

Figure 2.
Illustrations of flowfield over (a) cone (b) space vehicle at M_∞ = 3.1.

show several flowfield features such as free-shear layers, contraction of flow (neck) region and recompression shocks. The base flowfields also exhibit near and far wake region as depicted in **Figure 1(b)**. The values of L_c and h as depicted in **Figure 1(a)** are function several flow variables as mentioned above. The base plane of the capsule experiences another stagnation point.

Figure 2(a) and **(b)** has been drawn with the help of shadowgraph pictures of a 12.5° semi-cone and a blunt body capsule at high speed. The base pressure profile is illustrated in the wake region of the space vehicles. The schematic sketches as shown in **Figure 2** delineate a complex flowfield features associated with the nonlinear base pressure variations in the wake region.

4. Geometrical parameters of reentry vehicles

A high-speed flow past a reentry capsule forms a bow shock wave which causes a high surface pressure. It yields high aerodynamic drag (ballistic coefficient) force, which is needed for aero-braking purposes. Therefore, the primary design consideration of the reentry capsules requires large spherical nose radius R_N and fore-body diameter D as shown in **Figure 1(a)**. Reentry capsule configurations significantly differ from each other due to entry conditions and mission requirements. The sphere space capsule (Sputnik) permits the highest possible volumetric efficiency but does not give good maneuvering ability. Therefore, the reentry space vehicle requires a back-shell with an inclination in order to generate lift to reduce 'g' forces on the crew tolerance levels. Bedin et al. [52] have illustrated sixteen types of space vehicles in which the frontal diameter D of the capsule is kept constant for all configurations and varying geometrical parameters α_N, R_C, α_B, and L in three groups. Experimental investigation of various combination of cone-segment bodies and spheres of Russian reentry capsules are carried out by Bedin et al. [52] in a pressure-tight ballistic range for ratio of specific heats of 1.14–1.67, Mach number varied from 0.5 to 10, and Reynolds number based on the base diameter varied from 2.5×10^5 to 5.0×10^6.

In the first group, five capsules are having variation in the back-shell angle α_B in the range of 0–30°. In the second group, five capsules are having variation of the overall length varied from 1.0 D to 0.375 D. In the last group the back-shell angle α_B, overall length L, and shoulder radius R_C, alignment with frontal cap are varied to evaluate the ballistic performance. Recently Minenkol et al. [53] have studied the effect of geometrical parameters on aerodynamic performance of the space vehicles such as the Apollo and the Soyuz.

The reentry capsules can be classified as a head-light shape as in the case of Soyuz, or bell shape as in the case of Apollo and ARD, or a saucer type as in the case of OREX. **Table 1** depicts the dimension of the Apollo, the OREX and the Soyuz capsules to emphasis the classification of the capsules based on L/D ratio. The nominal OCM geometry, based on the Apollo configuration, consists of a spherical fore-body transitioning to a conical back-shell section with a truncated base to accommodate docking hardware. The aerodynamic characteristic of the Orion is analyzed numerically and experimentally by Stremel et al. [54]. The OCM is similar in shape to the Apollo Command Module but is approximately 29% larger by length. The ARD resembles a 70% scaled version of Apollo capsule as mentioned by Walpot [6].

The schematic sketches of flowfield feature of the Apollo, the Soyuz, the OREX capsules are displayed in **Figure 3(a)–(c)**. The Apollo and the Soyuz configurations are having spherical-blunt nose segment. The fore-body of the OREX consists of spherical cap with a cone section. The bow shock wave is detached on the blunt fore-body in the case of SRE as delineated in **Figure 3(d)**. The fore-body of the SRE is having a mixed subsonic-supersonic region as seen in the figures. The flowfield in the wake region is affected due to the presence of the truncated cylinder. **Figure 3(e)** shows schematic flowfield features at high speed on a sharp-tipped double-cone configuration. The double cone capsule shows formation of an attached conical shock wave on the tip of the cone. The flowfield in the wake region of a reentry capsule is again found to be complex in nature and is attributed to the expansion fan at corner of the shoulder.

Figure 4 shows the nomenclature of the geometrical parameters of the ARD, the Soyuz, the OREX, the SRE and the double cone reentry capsules. The Soyuz, the Apollo and the OREX capsules are having back-shell inclination angle α_B of 9, 33 and 15° relative to the vehicle's axis of symmetry respectively. **Figure 5** depicts the geometrical details of the CARINA [13] and Beagle-2 [14] capsules. **Table 2** depicts the geometrical detail of Viking, MPF, MER, Phoenix and MSL which are having a 70° sphere-cone shaped (Mars space vehicles) with a back-shell needed for high-speed entry phase and a disk-gap band (DGB) type of supersonic parachute during the descent portion of the entry sequence. **Table 3** presents the dimensional details of the ARD, the Apollo, the OREX, the CARINA, the MUSES-C and the Beagle-2. **Table 4** depicts the dimensional details of the SRE capsule.

Capsule	R_N	D	R_C	L	$\alpha_N{}^0$	$\alpha_B{}^0$
Apollo	4.595	3.95	0.186	2.04	—	33.0
OREX	1.35	3.40	0.001	1.508	50.0	15.0
Soyuz	2.235	2.2	0.014	2.142	—	7.0

Table 1.
Dimension of the Apollo, the OREX and the Soyuz reentry capsules.

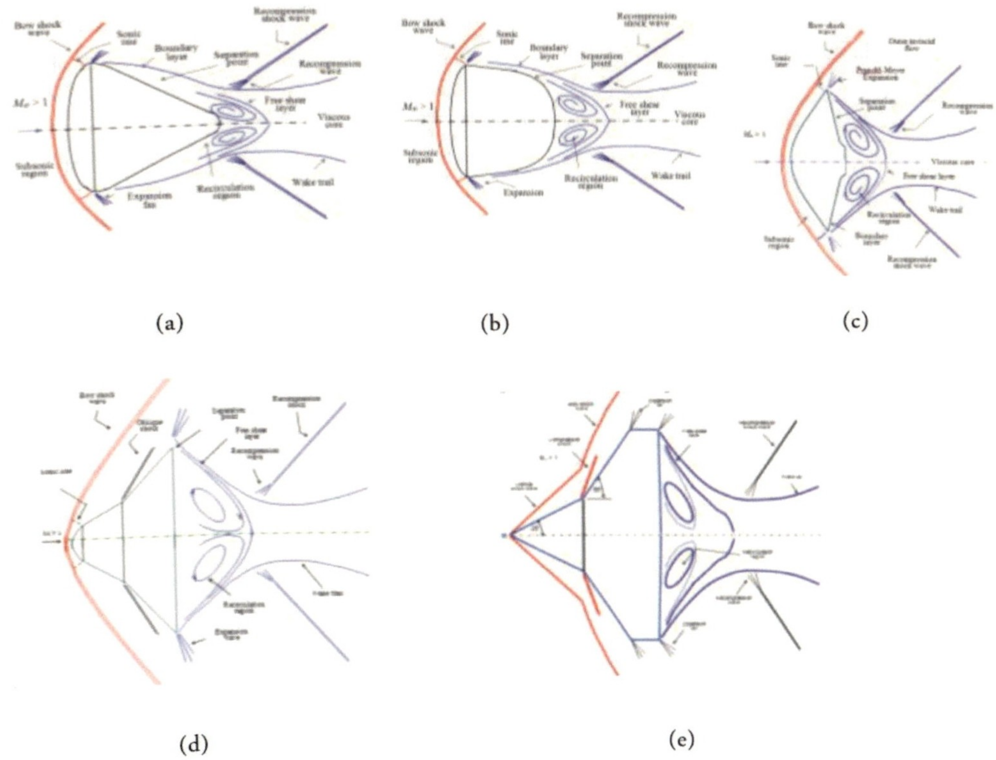

Figure 3.
Schematic sketches of flowfield over various reentry capsules (a) Apollo; (b) Soyuz; (c) OREX; (d) SRE; and (e) double-cone capsule.

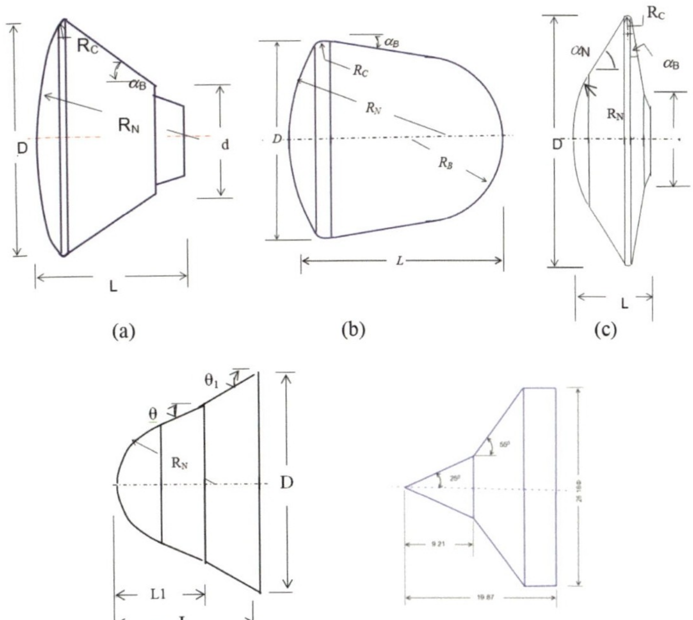

Figure 4.
Geometrical parameters of reentry capsules, (a) ARD; (b) Soyuz; (c) OREX; (d) SRE; and (e) double-cone space vehicles.

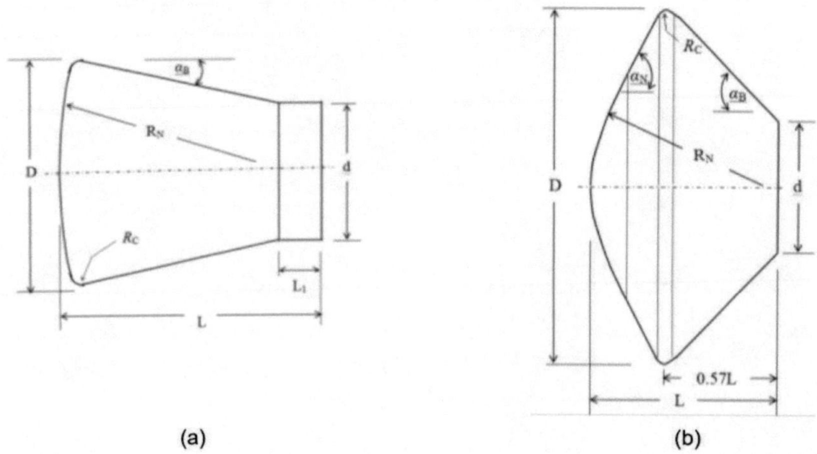

Figure 5.
Geometrical parameters of reentry capsules (a) CARINA; (b) Beagle-2.

Geometrical parameters	Capsules				
	Viking	MPF	MER	Phoenix	MSL
			$\alpha_N = 30°$		
Fore-body diameter, D	3.5	2.65	2.65	2.65	4.5
DGB parachute diameter	16.4	12.4	15.09	11.5	19.7

Table 2.
Geometrical parameters of Viking, MPF, MER, Phoenix and MSL.

Capsule	R_N	D	R_C	L	$\alpha_N{}^0$	$\alpha_B{}^0$
ARD	3.36	2.80	0.014	2.04	—	33
Apollo-II	4.595	3.95	0.186	3.52	—	33
OREX	1.35	3.40	0.001	1.50	50	15
CARINA	1.97D	1.0D	0.25D	1.172D	—	13
MUSES-C	2.0	4.0	—	2.0	45	45
Beagle-2	41.7	90.0	0.029	49.95	60	43.75

Table 3.
Dimension of the reentry capsules.

Semi-cone angle	R_N	D	L_1	L
$\theta = 25°$	3.36	2.80	0.014	2.04
$\theta = 30°$	4.595	3.95	0.186	1.50
$\theta = 35°$	1.35	3.40	0.001	1.50

Table 4.
Dimension of the blunted-spherical cone (SRE) reentry module.

The effects of the module geometrical parameters, such as radius of the spherical cap radius, shoulder radius, semi-cone angle and back-shell inclination angle on the flowfield characteristics hence the base drag coefficient are analyzed which will provide a useful input for the optimization of the reentry module.

5. Numerical algorithm

5.1 Governing fluid equations

As discussed above the base pressure measurements in the wind-tunnel testing are affected by presence of sting attached to model. The free-flight data depend on quality of the transmitted telemetry data. The fluid dynamic equations describing the flowfield around a space vehicle include equations of continuity, momentum, and total energy. A numerical simulation of unsteady, compressible, axisymmetric laminar Navier-Stokes equations is an alternative to the expensive experimental testing of the reentry vehicles. The governing fluid dynamics equations can be written in the following conservation form in order to capture shocks and discontinuities as

$$\frac{\partial \boldsymbol{U}}{\partial t}+\frac{\partial \boldsymbol{F}}{\partial x}+\frac{1}{r}\frac{\partial (r\boldsymbol{G})}{\partial r}+\frac{\boldsymbol{H}}{r}=0 \tag{3}$$

Temperature T is related to pressure and density by the perfect gas equation of state. The ratio of the specific heats γ is assumed constant and is equal to 1.4. The coefficient of molecular viscosity is evaluated in the flow solver employing Sutherland's formula. The flow is assumed to be laminar, which is consistent with experimental results of Cassanto [37] and Bulmer [42].

5.2 Numerical technique

To simplify the spatial discretization in numerical technique, Eq. (3) can be written in the integral form over a finite computational domain Ω with the boundary of the domain Γ as

$$\frac{d}{dt}\int_{\Omega}\boldsymbol{U}d\Omega+\int_{\Gamma}(\boldsymbol{F}dr-\boldsymbol{G}dx)+\int_{\Omega}\boldsymbol{H}d\Omega=0 \tag{4}$$

The contour integration around the boundary of the cell is performed in anticlockwise sense in order to keep flux vectors normal to boundary of the cell. The computational domain Ω is having a finite number of non-overlapping quadrilateral cells. The conservation variables within the computational cell are represented by their average values at the cell centre.

The inviscid fluxes are computed at the centre of the cell resulting in flux balance. The summation is carried out over the four edges of the cell. The derivatives of primitive variables in the viscous flux are evaluated by using the method of lines. A system of ordinary differential equations in time is obtained after integrating Eq. (4) over a computational cell. In the cell-centered spatial discretization scheme is non-dissipative, therefore, artificial dissipation terms [55] are added by blending of second and fourth differences of the vector conserved variables. The blend of second and fourth differences provides third order back ground dissipation in smooth region of the flow and first-order dissipation in shock waves.

The spatial discretization described above reduces the integral equations to semi-discrete ordinary differential equations (ODE). The ODE is solved using multi-stage Runge-Kutta time stepping scheme of Jameson et al. [55]. The numerical algorithm is second-order accurate in space discretization and time integration.

The scheme is stable for a Courant number ≤2. Local time steps are used to accelerate to a steady-state solution by setting the time step at each point to the maximum value allowed by the local Courant-Friedrichs-Lewy (CFL) condition.

5.3 Initial and boundary conditions

The freestream conditions for each trajectory point are tabulated in **Table 5**, which are used as initial conditions. The freestream flow values are used to initialize the whole flowfield.

The boundary conditions are as follows: a no-slip condition and isothermal wall is considered as a solid wall boundary condition. At the inflow, all the flow variables are taken at the freestream values as tabulated in **Table 5**. A symmetry condition is imposed on the centre line upstream and downstream of the reentry vehicle. All variables are extrapolated at the outer computational boundary.

5.4 Computational grid

The body oriented grids are generated using a homotopy scheme. The stretched grids are generated in an orderly manner. The grid-stretching factor is selected as 5,

M_∞	p_∞, Pa	T_∞, K
1.2	4519	210
1.4	3952	213
2.0	2891	219
3.0	2073	224
5.0	1238	232
6.0	1064	234

Table 5.
Trajectory points and initial conditions.

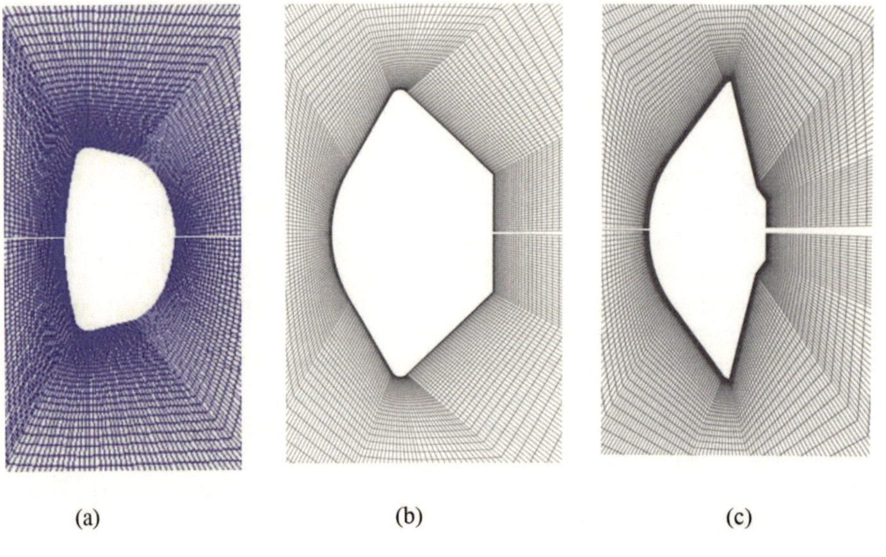

Figure 6.
Enlarged view of computational grid; (a) Soyuz; (b) MUSES-C; and (c) OREX.

and the outer boundary of the computational domain is maintained as 1.5–2.5 times maximum diameter D of the reentry module. In the downstream direction, the computational boundary is about 6–9 times the diameter of the module; D. **Figure 6** shows enlarged view of grid over the Soyuz, the MUSES-C and the OREX vehicle. The grid arrangement is found to yield a relative difference of about ±5% in the computation of fore-body aerodynamic drag coefficient. The convergence criterion is based on the difference in density values at any of the grid points, between two successive iterations $|\rho^{n+1} - \rho^{n}| \leq 10^{-5}$ where n is time-step counter. The present numerical algorithm is described in detail in Refs. [24, 25] and validated with many test cases.

6. Flowfield characteristics

Figure 7 depicts the velocity vector plots over the Apollo, the Apollo-II, the OREX and the MUSES-C space vehicles. It can be visualized from the vector plots that all the significant flowfield features such as a bow shock wave, rapid expansion fans at the shoulder, recirculation region with a converging free-shear layer and formation of the vortex flow in the base-shell region are well captured for $M_{\infty} = 5.0$. The wake flowfield immediately behind the space vehicle base exhibits complex flow characteristics. The formation of the bow shock wave on the fore-body

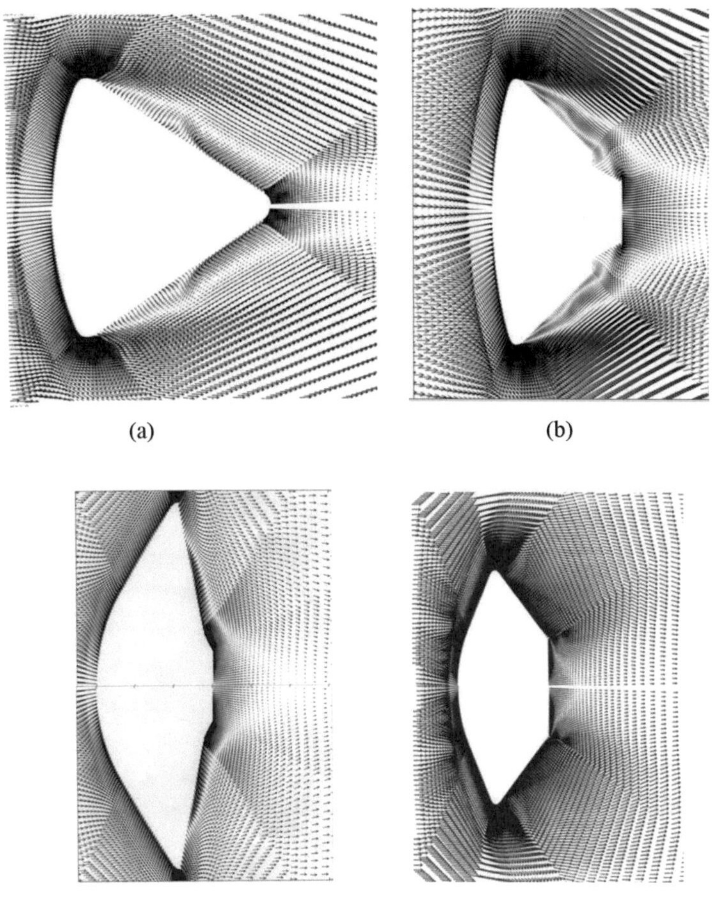

Figure 7.
Close-up views of velocity vector plots (a) Apollo; (b) ARD; (c) OREX and (d) MUSES-C at $M_{\infty} = 5.0$.

depends on R_N and α_N and M_∞. The bow shock wave moves close to the fore-body with the increasing M_∞ and the stand-off distance between the bow shock wave and the fore body decreases with the increasing M_∞. The Mach contour plots over the OREX and the MUSES-C are depicted in **Figure 8** for M_∞ = 1.2. The wake flowfield, immediately behind the capsule base, exhibits complex flow characteristics as observed in the vector plots.

Mach contours over the CARINA and Beagle-2 modules are exhibited in **Figure 9** for M_∞ = 1.2. The Mach contours over the SRE capsule for θ = 25° at M_∞ = 2.0 and 3.0 are shown in **Figure 10**. The bow shock wave does not follow the fore-body contour, which is due to small value of R_N and presence of semi-cone angle θ as compared to the OREX, the MUSES-C and the Apollo.

Figure 11(a) and **(b)** depicts velocity vector and Mach contour plots, respectively, over the double-cone (25/55°) configuration at M_∞ = 3. Despite its geometric simplicity, the double-cone shows the complex flowfield characteristics. A separation bubble can be observed on the vector plots. The separation and reattachment points are marked with the symbols "S" and"R"in the vector plots. It can also be seen from the vector plots that all the significant flowfield features are well captured such as the formation of conical shock wave on the tip, rapid expansion fan on the corner, recirculation region with converging free-shear layer and formation of the vortex flow in the aft region of the sharp-tipped double cone configuration.

The above numerical simulations over various reentry space capsules show that the separated flow can be found in the base region of the reentry capsules. The flow around the capsule is divided into two regions; inside and outside of the

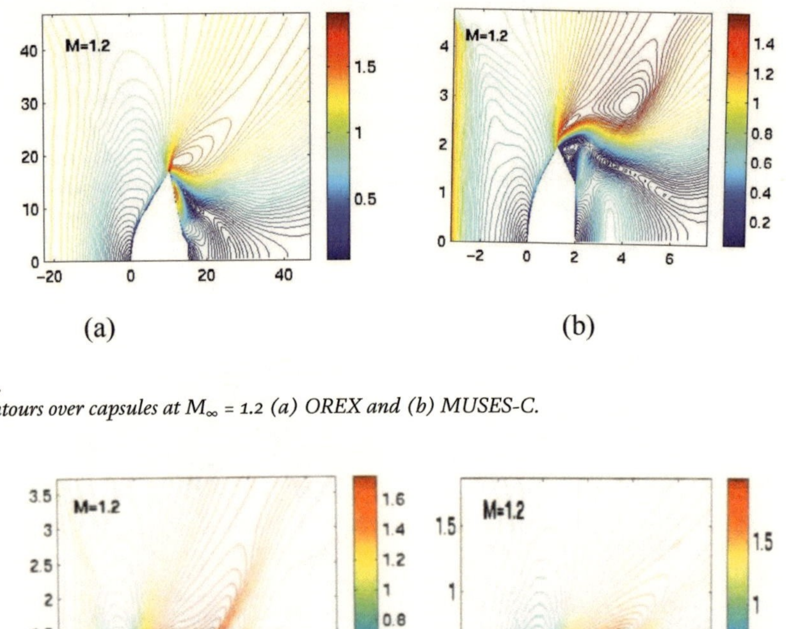

Figure 8.
Mach contours over capsules at M_∞ = 1.2 (a) OREX and (b) MUSES-C.

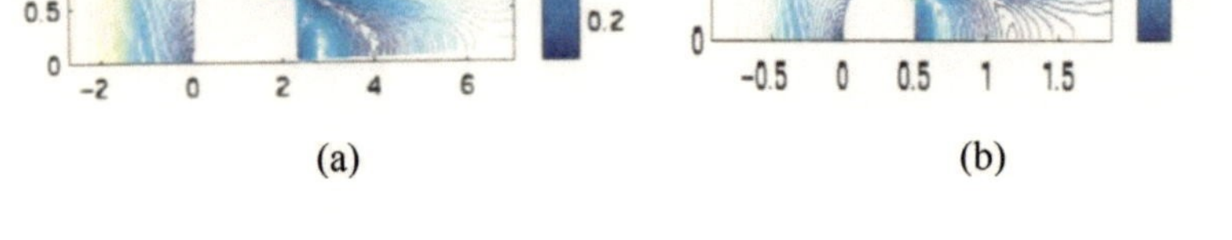

Figure 9.
Mach contours over (a) CARINA; and (b) Beagle-2 module at M_∞ = 1.2.

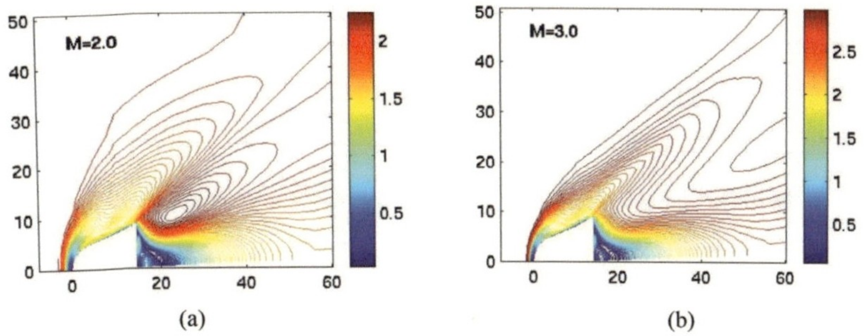

Figure 10.
Mach contour over SRE module at (a) M_∞ = 2.0 and (b) M_∞ = 3.0 at θ = 25°.

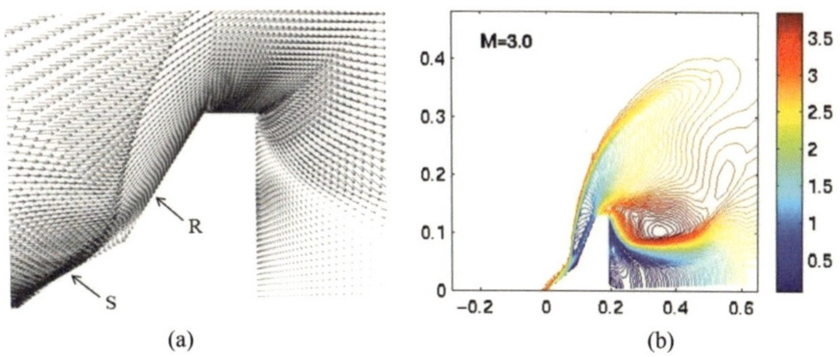

Figure 11.
(a) velocity vector and (b) Mach contours over double-cone module.

recirculation zone, and the shear layer separating the regions. The flowfield is very complex because of the back-shell. The wake flowfield, immediately behind the capsule base, exhibits vortex flow behavior. The formation of the bow shock wave on the fore body of the capsule depends on geometrical parameters such as spherical cap radius and the apex cone angle, and the value of the freestream Mach number. A low pressure is observed immediately downstream of the base which is characterized by a low-speed recirculating flow region, which can be attributed to filling-of the growing space between the shock wave and the reentry module. This flowfield behavior results the base drag.

Figure 12(a) and **(b)** depicts the variation of surface pressure coefficient Cp over the surface and the base plane, respectively, of the SRE capsule at M_∞ = 6.0, where s is measured along the surface of the fore-body. The s = 0 is the location of the stagnation point. The Cp variations is gradually decreasing over the spherical cap and remain nearly constant in the conical section of the SRE as depicted in **Figure 12(a)**. A sudden fall in Cp is seen on the sharp shoulder of the SRE. In the base region of the SRE, the Cp_B remains nearly linear variation on the base plane as seen in **Figure 12(b)**. The Cp_B is high on the corner due to presence of the expansion fan.

Figure 13(a) shows variation of Cp over the MUSES-C capsule at M_∞ = 3.0. A sudden drop in Cp is observed on the shoulder of the MUSES-C accompanied by a negative pressure coefficient Cp. The Cp_B is shown in **Figure 13(b)** for the MUSES-C space vehicle. The Cp_B remains near to a constant value on the base plane. It is important to mention here that the Cp_B variation is gradual attributed to beveled shape shoulder of the MUSES-C. Thus, the Cp and Cp_B variations over the SRE and the MUSES-C exhibit the influence of the geometrical parameters and freestream Mach number.

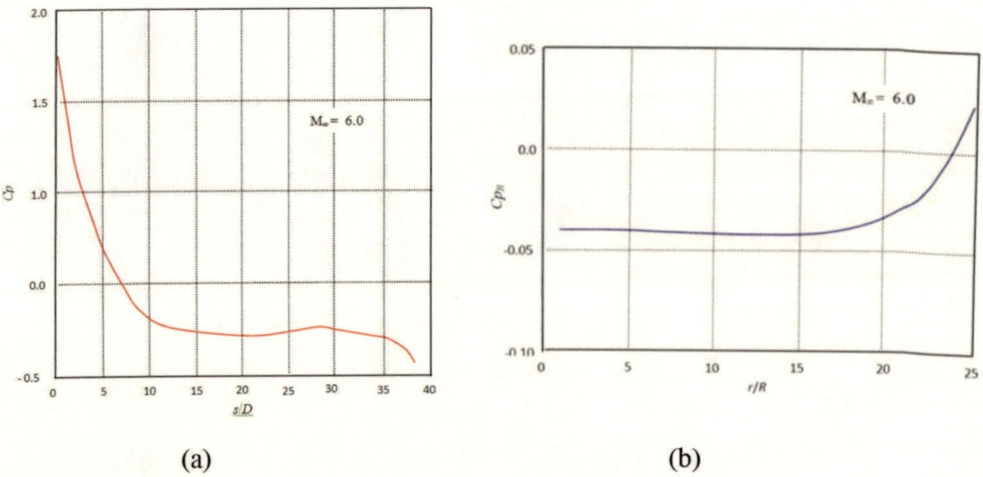

Figure 12.
Variation of pressure coefficient (a) over SRE module (b) on base region.

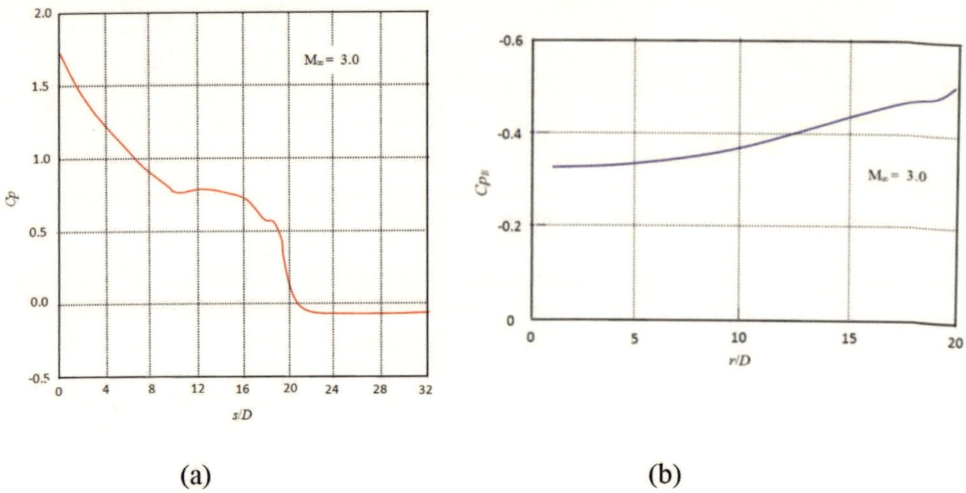

Figure 13.
Variation of pressure coefficient (a) over MUSES-C (b) on base region.

7. Base pressure and drag coefficients

Characteristics of flow features around the blunt body at supersonic speeds are described in the above section. The high surface pressure on the fore-body results in the high aerodynamic drag which is required for the aero-braking application. The base pressure coefficient can be calculated using following expression

$$C_{PBS} = \frac{(p_{BS} - p_\infty)}{\frac{1}{2}\rho_\infty V_\infty^2} \tag{5}$$

where subscript *BS* represents the base-stagnation point as depicted in **Figure 1(b)**. **Table 6** shows the computed base pressure coefficient C_{PBS} of the various capsules configurations at different freestream Mach numbers M_∞. **Table 6** shows OREX with smooth shoulder (beveled) and with a sharp corner. The C_{PBS} is high in the case of the OREX (S) as compared to the OREX with smooth shoulder. It again exhibits the effects of the shoulder shape geometry on the C_{PBS}.

Capsules	Cp_{BS}				
	M_∞ = 1.2	M_∞ = 2.0	M_∞ = 3.0	M_∞ = 5.0	M_∞ = 6.0
$Cp_{BS} = -2/(\gamma M_\infty^2)$	−0.9920	−0.7288	−0.3571	−0.1387	−0.0396
ARD	−0.5	−0.25	−0.15	−0.05	—
Soyuz	—	−0.50	−0.40	−0.20	—
Apollo	−0.30	−0.30	−0.20	−0.05	—
Apollo-II	−0.25	−0.30	−0.20	−0.08	—
OREX	−0.75	−0.30	−0.20	−0.10	—
OREX (S)	−0.90	−0.40	−0.25	−0.18	—
CARINA	−0.50	−0.30	−0.20	−0.05	—
MUSSES-C	−0.70	−0.30	−0.20	−0.10	—
Beagle-2	−0.8	−0.28	−0.15	−0.10	—
Double-cone, 25/55°	—	−0.25	−0.20	—	−0.05
SRE, θ = 25°	−0.82	−0.30	−0.20	—	−0.01
SRE, θ = 30°	−0.80	−0.32	−0.20	—	−0.01
SRE, θ = 35°	−0.70	−0.30	−0.20	—	−0.01

Table 6.
Pressure coefficient at base stagnation point of various reentry modules.

Capsule	C_{DB}				
	M_∞ = 1.2	M_∞ = 2.0	M_∞ = 3.0	M_∞ = 5.0	M_∞ = 6.0
OREX	-0.117×10^{-5}	-0.555×10^{-6}	-0.244×10^{-7}	-0.723×10^{-9}	
OREX (S)	-0.228×10^{-5}	-0.124×10^{-5}	-0.539×10^{-6}	-0.170×10^{-6}	
Carina	-0.389×10^{-4}	-0.649×10^{-5}	-0.978×10^{-5}	-0.162×10^{-5}	
Double cone	-0.268×10^{-3}	-0.606×10^{-4}	-0.337×10^{-4}	-0.200×10^{-3}	
MUSSES-C	-0.261×10^{-4}	-0.512×10^{-5}	-0.245×10^{-5}	-0.196×10^{-5}	
Beagle-2	-0.790×10^{-5}	-0.430×10^{-5}	-0.210×10^{-5}	-0.710×10^{-6}	
SRE θ = 20°	-0.261×10^{-4}	-0.331×10^{-4}	-0.146×10^{-4}		-0.467×10^{-5}
SRE θ = 25°	-0.517×10^{-5}	-0.159×10^{-4}	-0.360×10^{-4}		-0.851×10^{-4}
SRE θ = 30°	-0.622×10^{-5}	-0.254×10^{-4}	-0.111×10^{-4}		-0.355×10^{-4}
SRE θ = 35°	-0.383×10^{-6}	-0.138×10^{-4}	-0.318×10^{-4}		-0.177×10^{-4}

Table 7.
Base drag coefficient on various reentry capsules.

The aerodynamic drag is influenced by the fore-body shape. The fore-body aerodynamic drag coefficient for various reentry configurations at high speeds is earlier computed and tabulated in Ref. [56]. After body drag C_{DB} is calculated by integrating the surface pressure coefficient variation excluding the fore-body of the reentry vehicle and can be expressed as

$$C_{DB} = \frac{2\pi \int_i (C_{PB})\, r_i \sin\psi dx}{A_{\max}} \quad (6)$$

where r and ψ are local radius and local inclination angle in the x-direction station i respectively. A_{max} is the maximum cross-sectional area of the reentry module. **Table 7** shows the base body aerodynamic drag C_{DB} for various reentry modules. The present numerical simulation will be validated in future with experimentally measured data in order to assess the error bands between them. The influence of geometrical parameters of the space reentry capsules and freetream Mach number on the base pressure coefficient and the base drag coefficient can be seen in **Tables 6** and 7.

8. Conclusions

A main aim of the Chapter is to analyze numerically the base pressure over space reentry vehicles at freestream Mach number range of 1.2–6.0. A numerical algorithm is described to solve compressible laminar axisymmetric Navier-Stokes equations over various reentry capsules. The flowfield over the capsule reveals the effect of the geometrical parameters on the base pressure and base drag coefficients. The CFD methods yield flowfields over space vehicles without the interference of the sting-model attachment in wind tunnel experiments. A low pressure is formed in the base region of the capsule which is characterized by a low-speed recirculation region which can be due to fill-up the growing space. The approaching boundary layer separates at the corner and the free-shear layer is formed in the wake region. The wake flow also shows a vortex attached to the corner with a large recirculation, which depends on spherical nose radius, apex cone angle, back-shell inclination angle and freestream Mach number.

Nomenclature

C_D	drag coefficient
C_P	pressure coefficient
D	fore-body diameter
d	adapter diameter
$\mathbf{F}, \mathbf{G}$	flux vectors
$\mathbf{H}$	source vector
L	overall length
M	Mach number
p	static pressure
t	time
$\mathbf{U}$	conservative variables in vector form
R_N	radius of sphere
R_C	radius of shoulder
x, r	coordinate directions
α_N	semi-cone angle of fore-body
α_B	semi-cone angle of back-shell
γ	ratio of specific heats
θ	semi-cone angle
ρ	density
Subscripts	
B	base
BS	base stagnation point
∞	freestream condition

Author details

Rakhab C. Mehta[1,2]

1 Department of Aeronautical Engineering, Noorul Islam Center for Higher Education, India

2 School of Mechanical and Aerospace Engineering, Nanyang Technological University, Singapore

*Address all correspondence to: drrakhab.mehta@gmail.com

References

[1] Cassanto JE. A base pressure experiment for determining the atmospheric pressure profile of planets. Journal of Spacecraft and Rockets. 1973;**10**(4):253-261

[2] Lamb JP, Oberkampf WL. Review and development of base pressure and base heating correlations in supersonic flow. Journal of Spacecraft and Rockets. 1995;**32**(1):8-22

[3] Kawecki EJ. Comparison of several reentry capsule base pressure correlation. Journal of Spacecraft and Rockets. 1977;**14**(5):284-289

[4] Terry J, Barber T. CFD and experimental study of an inflatable reentry vehicle model at Mach 3 conditions. Acta Astronautica. 2007;**61**:854-865

[5] Togiti V, Lüdeke H. Computation of supersonic base flow using detached eddy simulation. In: Tropea C, Jakirlic S, Heinemann HJ, Henke R, Hönlinger H, editors. New Results in Numerical and Experimental Fluid Mechanics VI. Notes on Numerical Fluid Mechanics and Multidisciplinary Design (NNFM). Berlin, Heidelberg, Germany: Springer; 2007. p. 96

[6] Walpot I. Numerical analysis of the ARD capsule in S4 wind tunnel. In: Proceedings of the 4th European Symposium on Aerothermodynamics for Space Applications. Capua, Italy: ESA; 2001. pp. 523-527

[7] Chester W. Supersonic flow past a bluff body with a detached shock. Journal of Fluid Mechanics. 1965;**1**:353-365

[8] Freeman NC. On the theory of hypersonic flow past plane and axially symmetric bluff bodies. Journal of Fluid Mechanics. 1956;**1**:366-387

[9] Lighthill MJ. Dynamics of a dissociating gas, part 1: Equilibrium flow. Journal of Fluid Mechanics. 1957;**2**:1-32

[10] Wood AW, Gnoffo PA, Rault DFG. Aerodynamic analysis of commercial experiment transport reentry capsule. Journal of Spacecraft and Rockets. 1996;**33**(5):643-646

[11] Yamamoto Y, Yoshioka M. CFD and FEM Coupling Analysis of OREX Aero-Thermodynamic Flight Data, AIAA 95-2087, 1995

[12] Ivanov NM Catalogue of Different Shapes for Un-Winged Reentry Vehicles, Final Report, ESA Contract 10756/94/F/BM, 1994

[13] Solazzo MA, Sansome A, Gasbarri P. Aerodynamic characterization of the Carina reentry module in the low supersonic regimes. In: Proceedings of the 2nd European Symposium on Aerothermodynamics for Space Vehicles. Noordwiik, The Netherlands: ESTEC; 1994. pp. 41-47

[14] Liever PA, Habchi SD, Burnell SI, Lingard JS. Computational fluid dynamics prediction of the beagle 2 aerodynamic data base. Journal of Spacecraft and Rockets. 2003;**40**(5):632-638

[15] Mehta RC. High speed flowfield analysis for satellite launch vehicle and reentry capsule. Journal of Magnetohydrodynamics, Plasma and Space Research. 2008;**15**(1):51-99

[16] Ross JC, Brauckmann GJ. Aerodynamic and Aeroacoustic Wind Tunnel Testing of the Orion Space Craft, AIAA 2011-3160, 2011

[17] Murphy KJ, Bibb KL, Brauckmann GJ, Rhode MN, Owens B, Chan DT, et al. Orion crew module aerodynamic testing. In: 29th AIAA Applied Aerodynamics Conference; 27-30 June

2011; Honolulu, Hawaii, USA, AIAA Paper 2011-3502. 2011

[18] Zhenmiz Z, Yunliancy D, Yi L, Tieliang Z. Shape optimization design method for the conceptual design of reentry vehicles. Acta Aeronautica et Astronautica Sinica. 2011;**32**(11):1971-1979

[19] Ali SA, Husain M, Qureshi MN. Effects of nose-bluntness ratio on aerodynamic performance for reentry vehicle. Journal of Space Technology. 2012;**1**(1):38-41

[20] Viviani A, Pezzella G. Aerodynamic and Aerothermodynamic Analysis of Space Mission Vehicles. Switzerland: Springer International Publishing A. G.; 2015

[21] Chen B, Zhan H, Zhou W. Aerodynamic design of a reentry capsule for high speed manned reentry. Acta Astronautica. 2015;**106**:160-169

[22] Weiland C. Aerodynamic Data of Space Vehicles. Germany: Springer-Verlag Berlin Heidelberg; 2014

[23] Mehta RC. Effect of geometrical parameters of reentry capsules over flowfield at high speed flow. Advances in Aircraft and Spacecraft Science. 2017;**4**(4):487-501

[24] Mehta RC. Numerical simulation of supersonic flow past reentry capsules. Shock Waves. 2006;**15**(1):31-41

[25] Mehta RC. Computational investigation of flow oscillation over reentry capsules. International Journal of Computational Fluid Dynamics. 2006;**15**(2):247-260

[26] Prabhu R, Stewart J, Theraja R. Shock Interaction Studies on a Circular Cylinder at Mach 16, AIAA Paper 90-0606, 1990

[27] MacLean M, Wadhams T, Holden M, Candler G. Integration of CFD and experiments in the CUBRC LENS shock tunnel facilities to understand the physics of hypersonic and hypervelocity flows. In: 4th Symposium on Integrating CFD and Experiments in Aerodynamics; 14-16 September 2009; Rhode-Saint-Genèse, Belgium: Von Karman Institute. 2009

[28] Wright MJ, Sinha K, Olejniczak J, Candler GV, Magruder TD, Smits AJ. Numerical and experimental investigation of double-cone shock interactions. AIAA Journal. 2000;**38**(12):2268-2276

[29] Cassanto JE, Rasmussen NS, Coats JD. Correlation of free-flight base pressure data for M =4 to M = 19. AIAA Journal. 1969;7(6):1154-1157

[30] Nieden PZ, Olivier H. Determination of atmospheric densities from reentry flight data. Journal of Spacecraft and Rockets. 2007;**44**(2):332-337

[31] Cassanto JE. Effects of cane angle and bluntness ratio on base pressure. AIAA Journal. 1965;**3**(12):2351-2352

[32] Cassanto JE. Base pressure results at M = 4 using free-flight and sting-supported models. AIAA Journal. 1968;**6**(7):1411-1414

[33] Noack RW, Lopez AR. Inviscid Flowfield Analysis of Complex Vehicles: Vol. 1 Description of Numerical Methods, Report SAND-0776/1; Sandia National Laboratory, 1988

[34] McWherter M, Noack RW, Oberkamf WL. Evaluation of boundary layer and parabolized Navier-stokes solutions for reentry vehicles. Journal of Spacecraft and Rockets. 1986;**23**(1):70-78

[35] Baum E, King HH, Denison MR. Recent studies of the laminar base flow region. AIAA Journal. 1964;**2**(9):1527-1534

[36] Cassanto JE. Ratio on base pressure. AIAA Journal. 1965;**3**(12):351-352

[37] Cassanto JE. Radial base-pressure gradients in laminar flow. AIAA Journal. 1967;**5**(12):2278-2279

[38] Cassanto JE, Mendelson RS. Local flow effects on base pressure. AIAA Journal. 1968;**6**(6):1182-1185

[39] Cassanto JE, Schiff J, Softley EJ. Base pressure measurements on slender cones with domed after bodies. AIAA Journal. 1969;**7**(8):1607-1609

[40] Cassanto JE. Full-scale flight test base pressure results for a blunt planetary entry probe configuration. Journal of Spacecraft and Rockets. 1971;**8**(9):996-998

[41] Zarin NE. Base pressure measurements on sharp and blunt 9° cones at Mach numbers from 3.50 to 9.20. AIAA Journal. 1966;**4**(4):743-745

[42] Bulmer BM. Flight-test base pressure measurements in turbulent flow. AIAA Journal. 1976;**14**(12):1783-1785

[43] Sammonds RI, Kruse RL. Viking Entry Vehicle Aerodynamics at M = 2 in Air and some Preliminary Test Data for Flight in CO_2 at M = 11, NASA TN D-7974, 1975

[44] Venkatapathy E, Palmer G, Prabhu DK. AFE Base Computations Including Base Heating Predictions, AIAA Paper 91-1372, 1991

[45] Gnoffo PA, Weilmuenster R, Braun RD, Cruz CL. Influence of sonic line location on Mars path finder probe aerothermodynamics. Journal of Spacecraft and Rockets. 1996;**33**(2):169-179

[46] Tam LT. LU-SGS Implicit Scheme for Reentry Flow Computation and Comparison with Aerodynamic Flight Data, AIAA 92-2671, 1992

[47] Menne S. Computation of non-winged vehicle aerodynamics in the low supersonic range. In: Proceedings of the 2nd European Symposium on Aerothermodynamics for Space Vehicles; 21-25 November 1994; Noordwiik, The Netherlands: ESTEC. 1994. pp. 73-78

[48] Otten HBA. Preliminary computational investigation on aerodynamic phenomena DELFT aerospace reentry test vehicle. In: The Proceedings of the 4th European Symposium on Aerothermodynamics for Allocations; 15-18 October 2001; Capua, Italy: ESA. 2001. pp. 207-213

[49] Savino R, Peterna D. Blunted cone-flare in hypersonic flow. Computers and Fluids. 2005;**34**(7):771-890

[50] Barnhardt MD. Modeling and simulation of high speed wake flows [Ph.D. dissertation]. USA: University of Minnesota; 2009

[51] Suzuki K, Abe T. Transonic, Supersonic and Hypersonic Wind-Tunnel Tests on Aerodynamic Characteristics of Reentry Body with Blunted Cone Configuration, Report No. 658; Kanagawa, Japan: The Institute of Space and Aeronautical Science, 1995

[52] Bedin AP, Mishin GI, Chistyakove MV. Experimental investigation of the aerodynamic characteristics and geometric parameters of flows about various molecular structures. In: Koptev YI, editor. Gas Dynamics. NY, USA: Nova Science Publishers; 1992. pp. 39-67

[53] Minenkol VE, Agafonov DN, Yakushev AG. Project analysis of aerodynamics configuration of reentry capsule shaped body based on numerical methods for Newtonian flow theory. Aerospace Scientific

Journal. 2015;**4**:1-14. DOI: 10.7463/aersp.0415.0813899. In Russian

[54] Stremel PK, McMullen MS, Garcia JA. Computational Aerodynamic Simulations of the Orion Command Module, AIAA 2011-3503, 2011

[55] Jameson A, Schmidt W Turkel E. Numerical Solution of Euler Equations by Finite Volume Methods Using Runge-Kutta Time-Stepping Scheme, AIAA Paper 81-1259, 1981

[56] Mehta RC. Aerodynamic Drag Coefficient for Various Reentry Configurations at High Speed, AIAA 2006-3179, 2006